D0086802

Solid Matter

Joseph A. Angelo, Jr.

This book is dedicated to my wife, Joan.

SOLID MATTER

Facts On File, Inc.
An imprint of Infobase Learning
132 West 31st Street
New York NY 10001

Library of Congress Cataloging-in-Publication Data
Angelo, Joseph A.
Solid matter / Joseph A. Angelo, Jr.
p. cm.—(States of matter)
Includes bibliographical references and index.
ISBN 978-0-8160-7610-9
1. Solids—Popular works. 2. Solid state physics—Popular works. I. Title.
QC176.A56 2011
530.4'1—dc22 2010018626

Facts On File books are available at special discounts when purchased in bulk quantities for businesses, associations, institutions, or sales promotions. Please call our Special Sales Department in New York at (212) 967-8800 or (800) 322-8755.

You can find Facts On File on the World Wide Web at http://www.factsonfile.com

Text design by Annie O'Donnell
Composition by Hermitage Publishing Services
Illustrations by Sholto Ainslie
Photo research by the Author
Cover printed by Yurchak Printing, Inc., Landisville, Pa.
Book printed and bound by Yurchak Printing, Inc., Landisville, Pa.
Date printed: April 2011
Printed in the United States of America

10 9 8 7 6 5 4 3 2 1

This book is printed on acid-free paper.

Contents

Preface vi
Acknowledgments xi
Introduction xii

1 Solid Matter: An Initial Perspective 1
Basic Concept of Matter 1
The Notion of Atomism 5
Matter's Cosmic Heritage 9
Rankine—The Other Absolute Temperature 10
Mass-Energy 11
The Mystery of Dark Matter and Dark Energy 13
Taking a Really Close Look at Matter 16
The Standard Model 17

2 Physical Behavior of Matter 20
Understanding the Physical and Chemical Properties of Matter 20
Mass, Volume, and Density 21
Pressure 25
Temperature 26
Why Ice Floats 28
Chemical Elements Found on Earth 29
Chemical Bonds 32

3 The Gravity of Matter 35
The Notion of Gravity 35
Free-Falling Bodies 38
Newton's Mechanical Universe 40
Newton's Laws of Motion 42
Einstein's Novel Theory of Gravitation 43
Black Holes—Gravity's Ultimate Triumph over Matter 45

4 Fundamentals of Materials Science **50**
Basic Concepts 50
Materials Processing in Space 53
Elasticity 56
Some Important Concepts Associated with Heat 60
Electromagnetic Properties of Solid Matter 64
The First Chemical Battery 67
Optical Properties of Solid Matter 69

5 Rocks and Minerals **72**
The Solid Earth 72
The Rock Cycle 74
How Nature's Forces Created Yosemite National Park 77
Characteristics of Minerals 79
Salt—Pillar of Civilization 80
Minerals in Earth's Lithosphere 82
American History—Carved in Stone 85
Natural Gemstones 87

6 Metals **90**
General Characteristics of Metals 90
Corrosion 91
Gold's Electron Configuration 93
The Metals of Antiquity 96
Alloys and Superalloys 102
The Invention of Money 103
Gutenberg's Lead Alloy Revolutionizes the World 105
Metals and Modern Civilization 108
Famous Blackbird with Titanium Wings 109

7 Building Materials **112**
The Stone Age 112
Romans—Master Builders of the Ancient World 116
The Difference between Limestone and Marble 118
Dimension Stone and Crushed Stone 120
Modern Cement and Concrete 122

8 Carbon—Earth's Most Versatile Element 124
Carbon—General Characteristics and Applications 124
The Mystery of Diamonds and Graphite 125
World's Largest Diamond 128
Coal 129
Carbon Cycle 130
Carbon Sequestration 133
Fullerenes 134
Radiocarbon Dating 137

9 Sand, Silicon, and Ceramics 139
Sand 139
Silicon 141
Integrated Circuit (IC) 144
Silicon-Based Solar Cells 146
Ceramics 146

10 Polymers, Soft Matter, and Composites 150
Polymers 150
Amazing World of Plastics 152
Problem of Plastic Pollution 154
Soft Matter 155
Catching Comet Dust with Aerogel 157
Composite Materials 158

11 Conclusion 160

Appendix 164
Chronology 168
Glossary 179
Further Resources 196
Index 203

Preface

The unleashed power of the atom has changed everything save our modes of thinking.

—Albert Einstein

Humankind's global civilization relies upon a family of advanced technologies that allow people to perform clever manipulations of matter and energy in a variety of interesting ways. Contemporary matter manipulations hold out the promise of a golden era for humankind—an era in which most people are free from the threat of such natural perils as thirst, starvation, and disease. But matter manipulations, if performed unwisely or improperly on a large scale, can also have an apocalyptic impact. History is filled with stories of ancient societies that collapsed because local material resources were overexploited or unwisely used. In the extreme, any similar follies by people on a global scale during this century could imperil not only the human species but all life on Earth.

Despite the importance of intelligent stewardship of Earth's resources, many people lack sufficient appreciation for how matter influences their daily lives. The overarching goal of States of Matter is to explain the important role matter plays throughout the entire domain of nature—both here on Earth and everywhere in the universe. The comprehensive multivolume set is designed to raise and answer intriguing questions and to help readers understand matter in all its interesting states and forms—from common to exotic, from abundant to scarce, from here on Earth to the fringes of the observable universe.

The subject of matter is filled with intriguing mysteries and paradoxes. Take two highly flammable gases, hydrogen (H_2) and oxygen (O_2), carefully combine them, add a spark, and suddenly an exothermic reaction takes place yielding not only energy but also an interesting new substance called water (H_2O). Water is an excellent substance to quench a fire, but it is also an incredibly intriguing material that is necessary for all life here on Earth—and probably elsewhere in the universe.

Matter is all around us and involves everything tangible a person sees, feels, and touches. The flow of water throughout Earth's biosphere, the air people breathe, and the ground they stand on are examples of the most commonly encountered states of matter. This daily personal encounter with matter in its liquid, gaseous, and solid states has intrigued human beings from the dawn of history. One early line of inquiry concerning the science of matter (that is, *matter science*) resulted in the classic earth, air, water, and fire elemental philosophy of the ancient Greeks. This early theory of matter trickled down through history and essentially ruled Western thought until the Scientific Revolution.

It was not until the late 16th century and the start of the Scientific Revolution that the true nature of matter and its relationship with energy began to emerge. People started to quantify the properties of matter and to discover a series of interesting relationships through carefully performed and well-documented experiments. Speculation, philosophical conjecture, and alchemy gave way to the scientific method, with its organized investigation of the material world and natural phenomena.

Collectively, the story of this magnificent intellectual unfolding represents one of the great cultural legacies in human history—comparable to the control of fire and the invention of the alphabet. The intellectual curiosity and hard work of the early scientists throughout the Scientific Revolution set the human race on a trajectory of discovery, a trajectory that not only enabled today's global civilization but also opened up the entire universe to understanding and exploration.

In a curious historical paradox, most early peoples, including the ancient Greeks, knew a number of fundamental facts about matter (in its solid, liquid, and gaseous states), but these same peoples generally made surprisingly little scientific progress toward unraveling matter's inner mysteries. The art of metallurgy, for example, was developed some 4,000 to 5,000 years ago on an essentially trial-and-error basis, thrusting early civilizations around the Mediterranean Sea into first the Bronze Age and later the Iron Age. Better weapons (such as metal swords and shields) were the primary social catalyst for technical progress, yet the periodic table of chemical elements (of which metals represent the majority of entries) was not envisioned until the 19th century.

Starting in the late 16th century, inquisitive individuals, such as the Italian scientist Galileo Galilei, performed careful observations and measurements to support more organized inquiries into the workings of the natural world. As a consequence of these observations and experiments, the nature of matter became better understood and better quantified. Scientists introduced the concepts of density, pressure, and temperature in their efforts to more consistently describe matter on a large (or macroscopic) scale. As instruments improved, scientists were able to make even better measurements, and soon matter became more clearly understood on both a macroscopic and microscopic scale. Starting in the 20th century, scientists began to observe and measure the long-hidden inner nature of matter on the atomic and subatomic scales.

Actually, intellectual inquiry into the microscopic nature of matter has its roots in ancient Greece. Not all ancient Greek philosophers were content with the prevailing earth-air-water-fire model of matter. About 450 B.C.E., a Greek philosopher named Leucippus and his more well-known student Democritus introduced the notion that all matter is actually composed of tiny solid particles, which are *atomos* (ατομος), or indivisible. Unfortunately, this brilliant insight into the natural order of things lay essentially unnoticed for centuries. In the early 1800s, a British schoolteacher named John Dalton began tinkering with mixtures of gases and made the daring assumption that a chemical element consisted of identical indestructible atoms. His efforts revived atomism. Several years later, the Italian scientist Amedeo Avogadro announced a remarkable hypothesis, a bold postulation that paved the way for the atomic theory of chemistry. Although this hypothesis was not widely accepted until the second half of the 19th century, it helped set the stage for the spectacular revolution in matter science that started as the 19th century rolled into the 20th.

What lay ahead was not just the development of an atomistic kinetic theory of matter, but the experimental discovery of electrons, radioactivity, the nuclear atom, protons, neutrons, and quarks. Not to be outdone by the nuclear scientists, who explored nature on the minutest scale, astrophysicists began describing exotic states of matter on the grandest of cosmic scales. The notion of degenerate matter appeared as well as the hypothesis that supermassive black holes lurked at the centers of most large galaxies after devouring the masses of millions of stars. Today, cosmologists and astrophysicists describe matter as being clumped into enormous clusters and superclusters of galaxies. The quest for these scientists is to explain how the observable universe, consisting of understandable forms of matter and energy, is also immersed in and influenced by mysterious forms of matter and energy, called dark matter and dark energy, respectively.

The study of matter stretches from prehistoric obsidian tools to contemporary research efforts in nanotechnology. States of Matter provides 9th- to 12th-grade audiences with an exciting and unparalleled adventure into the physical realm and applications of matter. This journey in search of the meaning of substance ranges from everyday "touch, feel, and see" items (such as steel, talc, concrete, water, and air) to the tiny, invisible atoms, molecules, and subatomic particles that govern the behavior and physical characteristics of every element, compound, and mixture, not only here on Earth, but everywhere in the universe.

Today, scientists recognize several other states of matter in addition to the solid, liquid, and gas states known to exist since ancient times. These include very hot plasmas and extremely cold Bose-Einstein condensates. Scientists also study very exotic forms of matter, such as liquid helium (which behaves as a superfluid does), superconductors, and quark-gluon plasmas. Astronomers and astrophysicists refer to degenerate matter when they discuss white dwarf stars and neutron stars. Other unusual forms of matter under investigation include antimatter and dark matter. Perhaps most challenging of all for scientists in this century is to grasp the true nature of dark energy and understand how it influences all matter in the universe. Using the national science education standards for 9th- to 12th-grade readers as an overarching guide, the States of Matter set provides a clear, carefully selected, well-integrated, and enjoyable treatment of these interesting concepts and topics.

The overall study of matter contains a significant amount of important scientific information that should attract a wide range of 9th- to 12th-grade readers. The broad subject of matter embraces essentially all fields of modern science and engineering, from aerodynamics and astronomy, to medicine and biology, to transportation and power generation, to the operation of Earth's amazing biosphere, to cosmology and the explosive start and evolution of the universe. Paying close attention to national science education standards and content guidelines, the author has prepared each book as a well-integrated, progressive treatment of one major aspect of this exciting and complex subject. Owing to the comprehensive coverage, full-color illustrations, and numerous informative sidebars, teachers will find the States of Matter to be of enormous value in supporting their science and mathematics curricula.

Specifically, States of Matter is a multivolume set that presents the discovery and use of matter and all its intriguing properties within the context of science as inquiry. For example, the reader will learn how the ideal gas law (sometimes called the ideal gas equation of state) did not happen overnight. Rather, it evolved slowly and was based on the inquisitiveness and careful observations of many scientists whose work spanned a period of about 100 years. Similarly, the ancient Greeks were puzzled by the electrostatic behavior of certain matter. However, it took several millennia until the quantified nature of electric charge was recognized. While Nobel Prize–winning British physicist Sir J. J. (Joseph John) Thomson was inquiring about the fundamental nature of electric charge in 1898, he discovered the first subatomic particle, which he called the electron. His work helped transform the understanding of matter and shaped the modern world. States of Matter contains numerous other examples of science as inquiry, examples strategically sprinkled throughout each volume to show how scientists used puzzling questions to guide their inquiries, design experiments, use available technology and mathematics to collect data, and then formulate hypotheses and models to explain these data.

States of Matter is a set that treats all aspects of physical science, including the structure of atoms, the structure and properties of matter, the nature of chemical reactions, the behavior of matter in motion and when forces are applied, the mass-energy conservation principle, the role of thermodynamic properties such as internal energy and entropy (disorder principle), and how matter and energy interact on various scales and levels in the physical universe.

The set also introduces readers to some of the more important solids in today's global civilization (such as carbon, concrete, coal, gold, copper, salt, aluminum, and iron). Likewise, important liquids (such as water, oil, blood, and milk) are treated. In addition to air (the most commonly encountered gas here on Earth), the reader will discover the unusual properties and interesting applications of other important gases, such as hydrogen, oxygen, carbon dioxide, nitrogen, xenon, krypton, and helium.

Each volume within the States of Matter set includes an index, an appendix with the latest version of the periodic table, a chronology of notable events, a glossary of significant terms and concepts, a helpful list of Internet resources, and an array of historical and current print sources for further research. Based on the current principles and standards in teaching mathematics and science, the States of Matter set is essential for readers who require information on all major topics in the science and application of matter.

Acknowledgments

I wish to thank the public information and/or multimedia specialists at the U.S. Department of Energy (including those at DOE headquarters and at all the national laboratories), the U.S. Department of Defense (including the individual armed services: U.S. Air Force, U.S. Army, U.S. Marines, and U.S. Navy), the National Institute of Standards and Technology (NIST) within the U.S. Department of Commerce, the U.S. Department of Agriculture, the National Aeronautics and Space Administration (NASA) (including its centers and astronomical observatory facilities), the National Oceanic and Atmospheric Administration (NOAA) of the U.S. Department of Commerce, and the U.S. Geological Survey (USGS) within the U.S. Department of the Interior for the generous supply of technical information and illustrations used in the preparation of this book set. Also recognized here are the efforts of Frank Darmstadt and other members of the Facts On File team, whose careful attention to detail helped transform an interesting concept into a polished, publishable product. The continued support of two other special people must be mentioned here. The first individual is my longtime personal physician, Dr. Charles S. Stewart III, M.D., whose medical skills allowed me to successfully work on this interesting project. The second individual is my wife, Joan, who for the past 45 years has provided the loving and supportive home environment so essential for the successful completion of any undertaking in life.

Introduction

The history of civilization is essentially the story of the human understanding and manipulation of matter. This book presents many of the discoveries that led to the scientific interpretation of matter in the solid state. Readers will learn how the ability of human beings to relate the microscopic (atomic level) behavior of solid matter to readily observable macroscopic properties (such as density, hardness, elasticity, and strength) helped transform the world.

Supported by a generous quantity of full-color illustrations and interesting sidebars, *Solid Matter* introduces the reader to the basic characteristics and properties of solid matter. The three most familiar states of matter are solid, liquid, and gas. In general, a solid occupies a specific fixed volume and retains its shape. A liquid also occupies a specific volume but is free to flow and assume the shape of the portion of the container it occupies. A gas has neither a definite shape nor a specific volume. Rather, it will quickly fill the entire volume of a closed container. Unlike solids and liquids, gases can be compressed easily. At very high temperatures, another state of matter appears. This fourth state of matter is called plasma. Finally, as temperatures fall very low and approach absolute zero, scientists encounter a fifth state of matter called the Bose-Einstein condensate (BEC).

Throughout most of human history, understanding the true nature of matter has been a very gradual process. Although the concept of the atom originated in ancient Greece about 2,400 years ago, the majority of the important concepts, breakthrough experiments, and technical activities associated with the discovery and application of the atomic nucleus are actually less than a century old.

The story of matter starts with the beginning of the universe, an event called the big bang. *Solid Matter* briefly describes the cosmic connection of the elements and then leads the reader through several key events in human prehistory that resulted in more advanced uses of matter in the solid state. From prehistory to the present, the story of people parallels the human mind's restless search for the scientific meaning of substance. While science has unveiled some of matter's most intriguing mysteries, other aspects of matter remain strange and unexplained.

Solid Matter deals primarily with the characteristics and properties of so-called *ordinary* solid matter. According to modern scientists, ordinary matter is composed primarily of baryons. Baryons (such as protons and neutrons) are the heavy nuclear particles formed by the union of three quarks. In addition to ordinary matter, the universe also consists of the mysterious phenomena known as dark matter and dark energy.

The human use of solid materials began when a prehistoric ancestor of modern humans picked up a stone and used that stone as a tool or weapon. The Paleolithic period (Old Stone Age) represents the longest phase in human development. Scientists often divide this large expanse of time into three smaller periods: the Lower Paleolithic period, the Middle Paleolithic period, and the Upper Paleolithic period. Taken as a whole, the most important development that occurred during the Paleolithic period was the evolution of the human species from a near-human apelike creature into modern human beings (called *Homo sapiens*). The process was exceedingly slow, starting about 2 million years ago and ending about 10,000 years ago with the start of the Mesolithic period (Middle Stone Age)—an event coinciding with the end of the last ice age.

During the Lower Paleolithic period (approximately 2 million years ago to about 100,000 B.C.E.), early hunters and gatherers learned how to use simple stone tools for cutting and chopping. They also learned how to construct crude handheld stone axes. In the Middle Paleolithic period (approximately 100,000 B.C.E. to about 40,000 B.C.E.), Neanderthals lived in caves, learned to control fire, and used improved stone tools for hunting. These nomadic people also learned how to use bone needles to sew furs and animal skins into body coverings. During the Upper Paleolithic period (from 40,000 B.C.E. to approximately 10,000 B.C.E.), Cro-Magnons arrived on the scene and displaced Neanderthals. Cro-Magnon tribal clans enjoyed more efficient hunting and fishing activities. Improvements in tools and weapons included carefully sharpened obsidian and flint blades. Cro-Magnons had better sewn clothing, the first human-constructed shelters, and jewelry. Prehistoric jewelry included bone or ivory necklaces, natural gemstones, and even gold nuggets.

The Neolithic Revolution was the incredibly important transition from hunting and gathering to agriculture. As the last ice age ended about 12,000 years ago, various prehistoric societies in the Middle East (the Nile Valley and Fertile Crescent) and in parts of Asia independently began to adopt crop cultivation. Since farmers tend to stay in one place, the early peoples involved in the Neolithic Revolution began to establish semipermanent and then permanent settlements. During this period, people learned to use clay to make pottery and habitats. The development of pottery supported the storage and transport of food and drink, while the development of clay bricks enabled the construction of the first cities. Historians often identify this period as the beginning of human civilization.

Several thousand years ago, the use of metals and other materials supported the rise of advanced civilizations all around the Mediterranean Sea. Gold, copper, and silver were the first metals discovered and used by ancient peoples because these metals are sometimes found in nature in pure form. The discovery of alloys, such as bronze, further stimulated an amazing wave of progress. The Bronze Age and the Iron Age represent two very important materials science–driven milestones in human development. The Roman Empire served as an enormous political stimulus and evolutionary endpoint.

The rise of human civilization depended on the creative use of other solid materials in addition to metals. One crystalline solid, sodium chloride (NaCl), more commonly called salt, proved especially vital. Because salt could be used to preserve food, it was a highly valued material. By using salt, early peoples were able to extend their food supplies and sustain themselves between agricultural seasons. In addition, salt-preserved food allowed soldiers, explorers, and merchants to travel extended distances without dependence on local food supplies.

When the Roman Empire collapsed, western Europe fragmented into many small political entities, while eastern Europe, Asia Minor, and Northern Africa experienced a rising Muslim influence. Progress in materials science continued, but at a much more fractionated and sporadic pace—until the 17th century and the start of the scientific revolution in western Europe.

Prior to the Scientific Revolution, people in many early civilizations practiced an early form of chemistry known as alchemy. For most practitioners, alchemy was a mixture of mystical rituals, magical incantations, early attempts at materials science, and experimentation involving chemical reactions and procedures. Despite its often haphazard practices, medieval alchemy eventually gave rise to the science of chemistry in western Europe.

Early alchemists were influenced in their thinking by the four classic elements of ancient Greece, namely earth, air, water, and fire. Hellenistic-era peoples believed that all matter consisted of these four elements mixed in different proportions. Many alchemists thought that by changing the proportions of these basic elements they could change, or transmute, them into other substances. Influenced by alchemical practices from ancient Egypt, Hellenistic Greece, and the Arab world, medieval alchemists pursued two major objectives. First, they searched for something called the philosopher's stone—an object that could turn base metals into gold. Second, they pursued the elixir of life—a special potion that could confer immortality.

The initial breakthroughs in accurately understanding the behavior of solid matter took place during the scientific revolution. *Solid Matter* explores the pioneering efforts of key individuals whose discoveries and insights established the foundations of physics, chemistry, and other physical sciences during this period of intellectual awakening and enrichment. The scientific method continues to exert a dominant influence on human civilization. Historically aligned with the scientific revolution was the development of the steam engine and the major technosocial transformation called the *First Industrial Revolution.*

The chemical properties of matter remained a mystery until woven into an elegant intellectual tapestry called the periodic table. *Solid Matter* describes the origin, content, and application of the periodic table of the chemical elements. In 1869, the Russian chemist Dmitri Mendeleev (1834–1904) suggested that when organized according to their atomic weights in a certain row-and-column manner, the known chemical elements exhibited a periodic behavior of their chemical and physical properties. The appendix in this book contains a modern version of the periodic table with its greatly expanded contents. During the 20th century, the development of quantum mechanics and modern nuclear theory supported a more thorough understanding of the nature of matter and the periodic behavior of the elements.

Solid Matter describes how scientists explored the atomic nucleus and then discovered how to harvest the incredible amounts of energy hidden within. Scientific discoveries about the nuclear atom forever changed the trajectory of human civilization by enabling the nuclear age. This book also explains how the modern understanding of matter on the smallest, or quantum, scale emerged in the 20th century. Especially remarkable progress took place in the decades immediately following World War II. Theoretical physicists such as Murray Gell-Mann (1929–) examined the avalanche of data concerning fundamental particles and then suggested that supposedly elementary particles such as neutrons and protons actually had structure and smaller particles (called *quarks*) within them.

These activities supported the maturation of quantum mechanics and resulted in the emergence of the standard model of fundamental particles.

Solid Matter introduces the reader to some of the more interesting topics in materials science. Specific chapters address the physical and chemical behavior of matter, the intriguing relationship between matter and gravity, rocks and minerals, metals, building materials, the fascinating element carbon, silicon, and polymers. Special attention is also given to the role of semiconductor materials and the potential of nanotechnology. Nanotechnology is the on-going combined scientific and industrial effort involving the atomic-level manipulation of matter that promises to alter significantly the trajectory of civilization.

Solid Matter has been carefully designed to help any student or teacher who has an interest in the overall mysteries of matter: what it is, where it came from, how scientists measure and characterize it, and how knowledge of its fascinating properties and characteristics has shaped the course of human civilization. The back matter contains an index, periodic table appendix, chronology, glossary, and array of historical and current sources for further research. These should prove especially helpful for readers who need additional information on specific terms, topics, and events in humankind's continuing journey into the heart of matter.

The author has carefully prepared all the scientific material in this book so that a person familiar with SI units (the international language of science) will have no difficulty understanding and enjoying its contents. The author also recognizes that there is a continuing need for students and teachers in the United States to have units expressed in the American customary unit system. Consequently, where appropriate in the text, both unit systems appear side by side. An editorial decision places American customary units first, followed by the equivalent SI units in parentheses. This formatting decision in no way implies the author's preference for American customary units versus SI units. Rather, the author strongly encourages all readers to take advantage of this format to better learn and appreciate the important role that SI units play within the international scientific community.

CHAPTER 1

Solid Matter: An Initial Perspective

The ability of human beings to relate the microscopic (atomic level) behavior of solid matter to readily observable macroscopic properties (such as density, hardness, elasticity, ductility, and temperature) has transformed the world. The initial breakthroughs in accurately describing the various properties of matter took place during the scientific revolution. This chapter explores some of most important results of the pioneering efforts that established the foundations of physics, chemistry, and modern materials science during this important period of intellectual awakening and enrichment. The scientific method, developed as part of the Scientific Revolution, continues to exert a dominant influence on human civilization.

BASIC CONCEPT OF MATTER

Scientists characterize matter as anything that occupies space and has mass. Matter is the substance of all material objects in the universe. The three most common forms of matter found on Earth are solid, liquid, and gas. One important factor shared by all forms of matter is that mass gives rise to the important physical property called *inertia*. Inertia is the resistance of a body to a change in its state of motion. As first proposed by Sir Isaac Newton (1642–1727) in the 17th century, mass is the inherent property of a body that helps scientists quantify inertia. This book focuses upon the concept of inertia as it is related to solid matter.

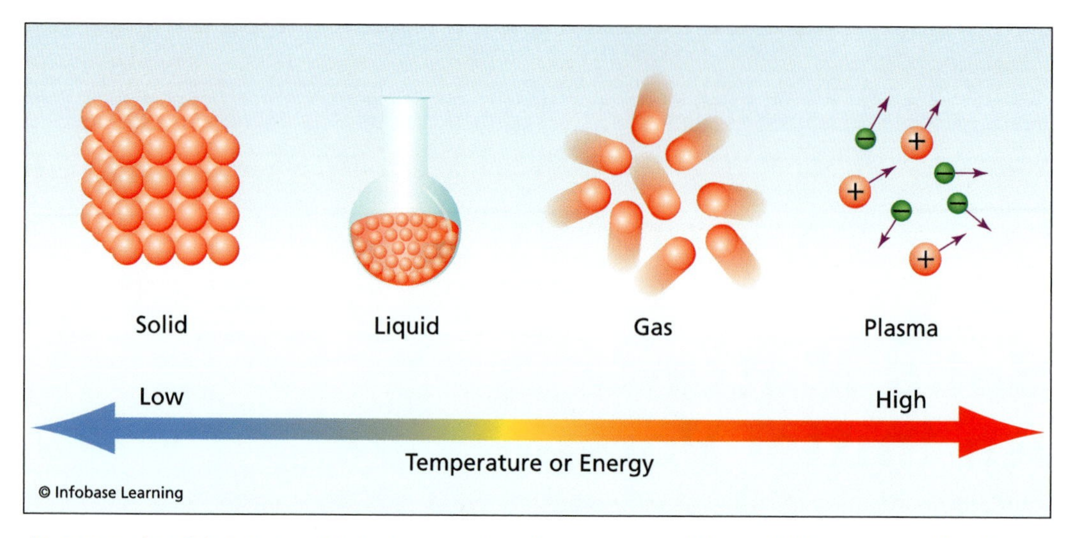

As energy is added to a solid, its temperature increases, and the solid becomes a liquid. Further addition of energy allows the liquid to become a gas. If the gas is heated a great deal more, its atoms break apart into charged particles, resulting in the fourth state of matter, plasma. *(Based on NASA-sponsored artwork)*

When examined at the macroscopic level, solid matter retains its shape, occupies a specific volume, and exhibits important bulk physical properties such as density, ductility, elasticity, strength, and hardness. When viewed at the microscopic (or atomic) level, the atoms or molecules of a solid are not very mobile and have a tendency to stay relatively fixed in their positions with respect to adjacent atoms or molecules. *Crystalline solids,* such as quartz, have atoms or molecules arranged in a very neat and orderly fashion. In contrast, *amorphous solids,* such as glass, have no large-scale, repetitive arrangement of their constituent atoms or molecules.

The addition of heat can cause a solid to melt, a change of state, and become a liquid. (The expression *phase change* is also commonly encountered during discussions of melting or evaporation.) On the macroscopic scale, the liquid form of a substance is quite different from the solid form of that substance. For example, a person can easily grasp a cube of ice with a pair of tongs but cannot use the same tool to pick up an equivalent quantity of liquid water.

A liquid assumes the shape of its container and occupies a specific volume within that container. Viewed at the atomic (or microscopic) level, the atoms or molecules that make up a liquid are free to move about much more than they can in a solid. Resembling miniature versions of shape-shifter creatures found in science fiction, the atoms or molecules of a liquid can slip and slide past each other and assume the shape of their container. But the forces between atoms or molecules in a liquid still exert considerable influence, which is why a liquid substance maintains a distinct, observable volume. When a scientist pours 2.1 quarts (qt) (2 liters [L]) of water from a bottle into a large bucket, the water will occupy a space at the bottom of the bucket equivalent to a volume of 2.1 qt (2 L).

If water in a flask is carefully heated, it will eventually boil and evaporate into another common form of matter called a gas. (Scientists often use the term *vapor* to describe a gas that readily condenses back to the liquid state.) At the macroscopic level, gases (including water vapor) do not keep their shape and do not maintain any definite volume. As a result, gases will simply occupy all the available space in a closed container. Gases will escape from an open container into the surrounding (lower pressure) environment until a uniform pressure condition, characterized by mechanical equilibrium, is achieved.

At the microscopic level, the atoms or molecules of a gas experience no significant interatomic or intermolecular forces. Rather, the atoms or molecules of a gas can travel relatively unhindered at high speed throughout

Rock crystal—a variety of the mineral quartz (SiO_2) *(NBII; photographer, Randolph Femmer)*

Foam is one type of soft matter. In this picture, two U.S. Air Force firefighters work their way through a sea of fire-suppression foam that had been automatically released in a hangar at Shaw Air Force Base, South Carolina. Because of the foam and quick human response, the fire caused only minimal damage within the hangar. *(U.S. Air Force)*

the entire enclosing volume until they physically collide with each other or the container wall. In the 19th century, scientists discovered that the average speed of atoms or molecules in a gas is related to the absolute temperature of the gas. Specifically, scientists observed that the higher the temperature of a gas, the higher the average speed of its constituent atoms or molecules.

Physicists call solids and liquids *condensed matter* and gases *uncondensed matter.* The term *soft matter* is a subfield within condensed matter physics and involves the study of condensed matter that is easily deformed. Examples of soft matter include elastomers, polymers, foams, and gels.

The property of matter called *mass* gives rise to the interesting physical phenomenon of inertia. Since electrically neutral matter is self-attractive, matter also gives rise to the very important force in nature call gravity (or gravitation). On Earth's surface, people commonly discuss how much mass an object has by expressing its *weight.* This is not scientifically correct, because the weight of an object is really a force—defined as the product of the local acceleration of gravity times a mass. (This originates from Newton's second law of motion, discussed in chapter 3.) The pioneering work of Galileo Galilei (1564–1642) and Newton in the 17th century led to the classical formulation of gravitation. Early in the 20th century, Albert Einstein (1879–1955) provided an expanded interpretation of gravitation within his general relativity theory. Gravitation is the important natural phenomenon that dominates the overall behavior of matter throughout the observable universe.

THE NOTION OF ATOMISM

Today, almost every student in high school or college has encountered the basic scientific theory that matter consists of atoms. This generally accepted model of matter was not always prevalent in human history. The notion of atomism traces its origins to ancient Greece.

The Greek philosopher Democritus (ca. 460–ca. 370 B.C.E.) was born in Abdera, Trace, in about 460 B.C.E. As a young man, he used his inherited wealth to travel the ancient world and then settled down in Thrace to focus on the practice of natural philosophy. Despite the intellectual contributions of his mentor Leucippus (fifth century B.C.E.), Democritus generally receives most of the credit for being the first person to promote atomism, the idea that an atom is the smallest piece of an element and indivisible by chemical means. Long before the emergence of the scientific method, Democritus reasoned that if a person continually divides a chunk of matter into progressively smaller pieces, he or she eventually reaches a point beyond which subdivision is no longer possible. At that point, only indivisible little building blocks of matter, or atoms, remain. The modern word *atom* comes from the ancient Greek word ατομος, meaning indivisible.

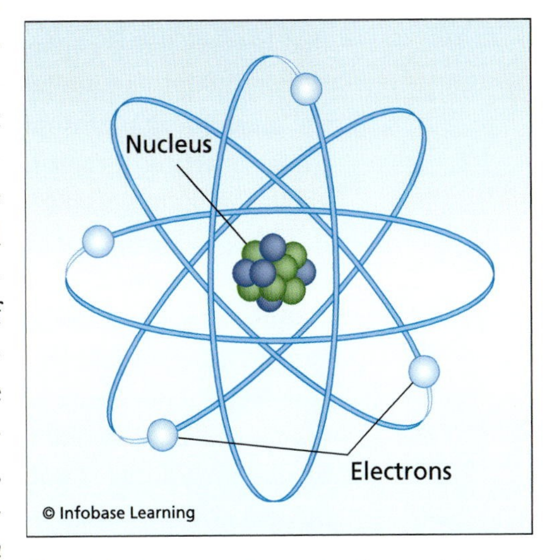

Simplified model of a typical (generic) atom with orbiting electrons shown in light blue, protons in the nucleus appearing in dark blue, and neutrons in the nucleus appearing in green (Illustration not drawn to scale) *(Based on DOE-sponsored artwork)*

Democritus proposed that these indivisible pieces of matter (or atoms) were eternal and could not be further divided or destroyed.

His atoms were also specific to the material they made up. Some types of solid matter consisted of atoms with hooks, so they could attach to each other. Other materials, such as water, consisted of large round atoms that moved smoothly past each other. What is remarkable about the ancient Greek theory of atomism is that it tried to explain the great diversity of matter found in nature with just a few basic ideas tucked into a relatively simple theoretical framework.

Despite Democritus's clever insight, the notion of the atom as the tiniest, indivisible piece of recognizable matter languished in the backwater of human thought for more than two millennia. The main reason for this intellectual neglect was the very influential Greek philosopher Aristotle (384–322 B.C.E.), who did not like the idea. Starting in about 340 B.C.E., Aristotle embraced and embellished the theory of matter originally proposed by Empedocles (ca. 495–435 B.C.E.). Within Aristotelian cosmology, planet Earth was regarded as the center of the universe. Aristotle speculated that everything within Earth's sphere was composed of a combination of the four basic elements: earth, air, water, and fire. Aristotle further suggested that objects made of these four basic elements were subject to change and moved in straight lines, but that heavenly bodies were not subject to change and moved in circles. Aristotle also proposed that beyond Earth's sphere lay a fifth basic element, which he called the aether (αιθηρ)—a pure form of air that was not subject to change. Finally, the great Greek philosopher suggested that he could analyze all material things in terms of their matter and their form (essence). Aristotle's ideas about the nature of matter and the structure of the universe dominated thinking in Europe for centuries until it was finally displaced during the scientific revolution.

Since Aristotle's teachings dominated Western civilization, the atomism of Leucippus and Democritus was all but abandoned. Another reason atomism did not flourish as an important concept linked to the nature of matter was the fact that the precision instruments and machines needed to effectively study atomic and nuclear phenomena began to appear only in the early part of the 20th century. Today, scientists have incredibly powerful machines that can image matter down to the atomic scale and measure intriguing processes that take place within the atomic nucleus, the very heart of matter.

During the 16th and 17th centuries, natural philosophers and scientists such as Galileo, René Descartes (1596–1650), Robert Boyle (1627–91), and Newton all favored the view that matter was not continuous in nature but

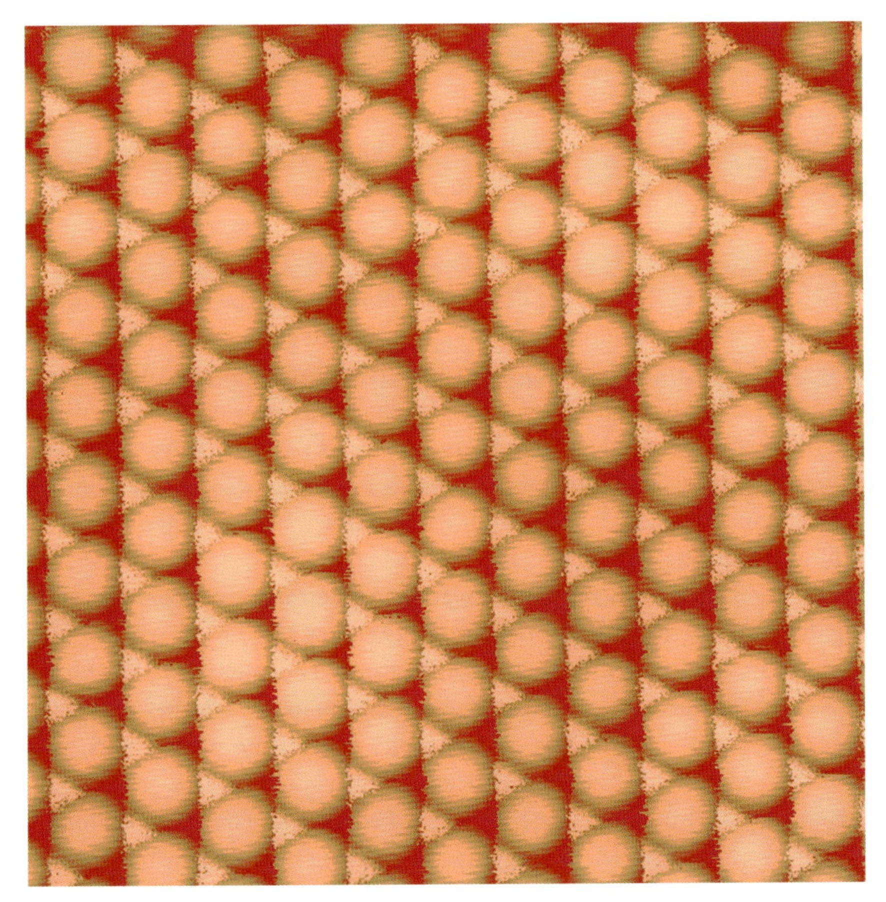

This amazing colorized image was created by the custom-built scanning tunneling microscope at the National Institute of Standards and Technology (NIST) as scientists dragged a cobalt atom across a closely packed lattice of copper atoms. *(Joseph Stroscio, Robert Celotta/NIST)*

rather consisted of discrete, tiny particles, or atoms. However, it was not until the 19th century and the hard work of several pioneering chemists and physicists that the concept of the atom was gradually transformed from a vague philosophical concept into a modern scientific reality.

In the 20th century, an exciting synergism occurred between the discovery of previously unimaginable nuclear phenomena and the emergence of new theories concerning the nature of matter and energy. Today, scientists use very powerful accelerators to experimentally probe inside nuclear particles. They hope to validate improved theories of matter and then relate these discoveries to contemporary observations in astrophysics, which suggest the existence of puzzling phenomena such as dark matter and dark energy.

Scientists recognize that an atom is the smallest particle of matter that retains its identity as a chemical element. Atoms are indivisible by chemical means and the fundamental building blocks of all matter. The chemical elements, such as hydrogen (H), helium (He), carbon (C), iron (Fe), lead (Pb), and uranium (U), differ from one another because they consist of different types of atoms. (See the appendix for a complete list of the known chemical elements.) Modern atomic theory suggests that an atom consists of a dense inner core (called the *nucleus*) that contains protons and neutrons and an encircling cloud of orbiting electrons. When atoms are electrically neutral, the number of positively charged protons equals the number of negatively charged electrons. The number of protons in an atom's nucleus determines what chemical element it is, while how an atom shares its negatively charged orbiting electrons determines the way that particular element physically behaves and chemically reacts. Through the phenomenon of covalent bonding, for example, an atom forms physically strong links by sharing one or more of its electrons with neighboring atoms. This atomic-scale linkage ultimately manifests itself in large-scale (that is, macroscopic) material properties, such as a substance's strength and hardness.

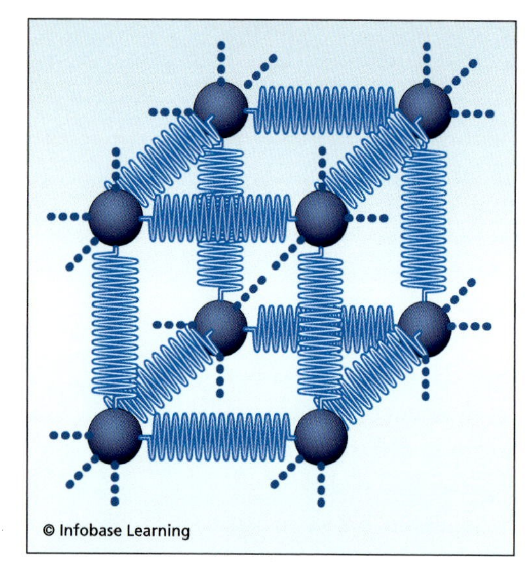

One convenient microscopic model of solid materials, such as crystals, is to assume that individual atoms are held together in orderly arrangements by stiff springs. *(Author)*

As pure substances, the minerals diamond and graphite each contain only carbon atoms. However, the carbon atoms are arranged in significantly different ways, thereby creating different molecular forms called *allotropes*. The carbon atoms in a diamond are held together by covalent bonds, which form a rigid three-dimensional crystal lattice that disperses light very well. Since every carbon atom in the diamond is tightly bonded to four other carbon atoms, the diamond has a rigid chain (or network) that results in the hardest known mineral. Graphite is another allotrope of carbon in which covalent bonds form sheets of atoms in hexagonal patterns, but the adjacent sheets of carbon atoms in graphite are more loosely bound to each other by much weaker intermolecular forces called van der

Waals forces. The impact of this particular molecular arrangement is startling, since graphite is also a crystalline mineral, but one that is soft and brittle. (See chapter 8.)

To help explain how the orderly arrangement of atoms in some solid materials determines macroscopic properties, scientists use a microscopic model in which each of the atoms is assumed to be connected to its neighbors by very tiny rigid springs. Within the model, the atoms in a crystalline lattice can vibrate but do not leave their basic positions relative to adjacent atoms. This model is quite useful in understanding some of the interesting properties of solid substances discussed later in the book.

MATTER'S COSMIC HERITAGE

Throughout most of human history, people considered themselves and the planet they lived on as being apart from the rest of the universe. After all, the heavens were clearly unreachable and therefore remained the celestial abode of the various deities found in the numerous mythologies that enriched ancient civilizations. It is only with the rise of modern science that people have been able to properly ascertain the cosmic heritage of the chemical elements as part of the overall evolution of the physical universe following the big bang event.

All the natural chemical elements found on Earth and elsewhere in the universe have their ultimate origins in cosmic events. Since different elements come from different events, the elements found on Earth—including those that make life itself possible—reflect an interesting variety of cosmic phenomena that have taken place in the universe over the past 13.7 billion years. The hydrogen found in water and hydrocarbon molecules was formed just a few moments after the big bang event that started the universe, but carbon, the element considered the basis for all terrestrial life, and other light elements, such as calcium and potassium, were formed by nucleosynthesis in the interior of stars. Heavier elements—those with atomic numbers beyond iron, such as silver, gold, thorium, and uranium—were formed by various neutron capture reactions deep in the interiors of highly evolved stars or during supernovas. Certain light elements, such as lithium, beryllium, and boron, resulted from energetic cosmic ray interactions with the atomic nuclei of hydrogen, helium, and other elements found in interstellar space.

Following the big bang explosion, the early universe contained the primordial mixture of energy and matter that evolved into all the forms

RANKINE—THE OTHER ABSOLUTE TEMPERATURE

Most of the world's scientists and engineers use the Kelvin scale (named after Lord Kelvin, who was also known as William Thomson [1824–1907]) to express absolute thermodynamic temperatures. However, there is another absolute temperature scale called the Rankine scale (symbol R) that sometimes appears in engineering analyses performed in the United States—analyses based upon the American customary system of units.

In 1859, the Scottish engineer and physicist William John Macquorn Rankine (1820–72) introduced the absolute temperature scale that now carries his name. Absolute zero in the Rankine temperature scale (that is, 0 R) corresponds to –459.67°F. The relationship between temperatures expressed in Rankines (R) using the (absolute) Rankine scale and those expressed in the degrees Fahrenheit (°F) using the (relative) Fahrenheit scale is T (R) = T (°F) + 459.67. The relationship between the Kelvin scale and the Rankine scale is (9/5) × absolute temperature (kelvins) = absolute temperature (rankines). For example, a temperature of 100 K is expressed as 180 R. The use of absolute temperatures is very important in science.

of energy and matter scientists observe in the universe today. About three minutes after the big bang, the temperature of this expanding mixture of matter and energy fell to approximately 1.8 billion rankines (R) (1×10^9 kelvins [K])—"cool" enough that neutrons and protons began to stick to each other during certain collisions and form light nuclei, such as deuterium, helium, and lithium. When the universe was three minutes old, about 95 percent of the atoms were hydrogen, approximately 5 percent were helium, and only trace amounts were lithium. At the time, these three elements were the only ones that existed.

As the universe continued to expand and cool, the early atoms (mostly hydrogen and lesser amounts of helium) slowly began to gather through gravitational attraction into very large clouds of gas. For millions of years, these giant gas clouds were the only matter in the universe, because neither stars nor planets had yet formed. Then, about 400 million years after the big bang, the first stars began to shine, and the creation of important new chemical elements started in their thermonuclear furnaces.

Gravitational attraction condensed portions of the giant clouds of primordial (big bang) hydrogen and helium into individual new stars. The very high temperatures in the cores of massive early stars supported the manufacture of heavier nuclei up to and including iron through a process called nucleosynthesis. Elements beyond iron were formed in a bit more spectacular fashion. Neutron capture processes deep inside highly evolved massive stars and subsequent supernova explosions at the end of their relatively short lifetimes synthesized all the elements beyond iron.

Astrophysicists explain that the slow neutron capture process (or *s-process*) produced heavy nuclei up to and including bismuth-209, the most massive naturally occurring nonradioactive nucleus. The flood of

MASS–ENERGY

In 1905, Einstein wrote "On the Electrodynamics of Moving Bodies." This paper introduced his special theory of relativity, which deals with the laws of physics as seen by observers moving relative to one another at constant velocity. Einstein stated the first postulate of special relativity as: "The speed of light *(c)* has the same value for all (inertial reference–frame) observers, independent and regardless of the motion of the light source or the observers." His second postulate of special relativity proclaimed: "All physical laws are the same for all observers moving at constant velocity with respect to each other."

From the special theory of relativity, Einstein concluded that only a zero rest mass particle, such as a photon, could travel at the speed of light. A major consequence of special relativity is the equivalence of mass and energy. Einstein's famous mass-energy formula, expressed as $E = mc^2$, provides the energy equivalent of matter and vice versa. Among its many important physical insights, this equation was the key that scientists needed to understand energy release in such important nuclear reactions as fission, fusion, radioactive decay, and matter-antimatter annihilation.

An electron has a mass of approximately 2.01×10^{-30} lbm (9.11×10^{-31} kg). The energy equivalent of an electron at rest is about 511 keV, which corresponds to 7.76×10^{-17} British thermal unit (Btu) (8.19×10^{-14} joules [J]). As a point of reference, an electron volt (eV) has the energy equivalent of 1.519×10^{-22} Btu (1.60×10^{-19} J). Now consider the mass-energy of something much larger than an electron, such as a U.S. one-cent coin. A typical American one-cent coin has a mass of about 6.62×10^{-3} lbm (3×10^{-3} kg [3 g]). If all the matter in just one penny were suddenly converted into pure energy, the amount of energy released would be quite large, about 2.56×10^{11} Btu (2.7×10^{14} J).

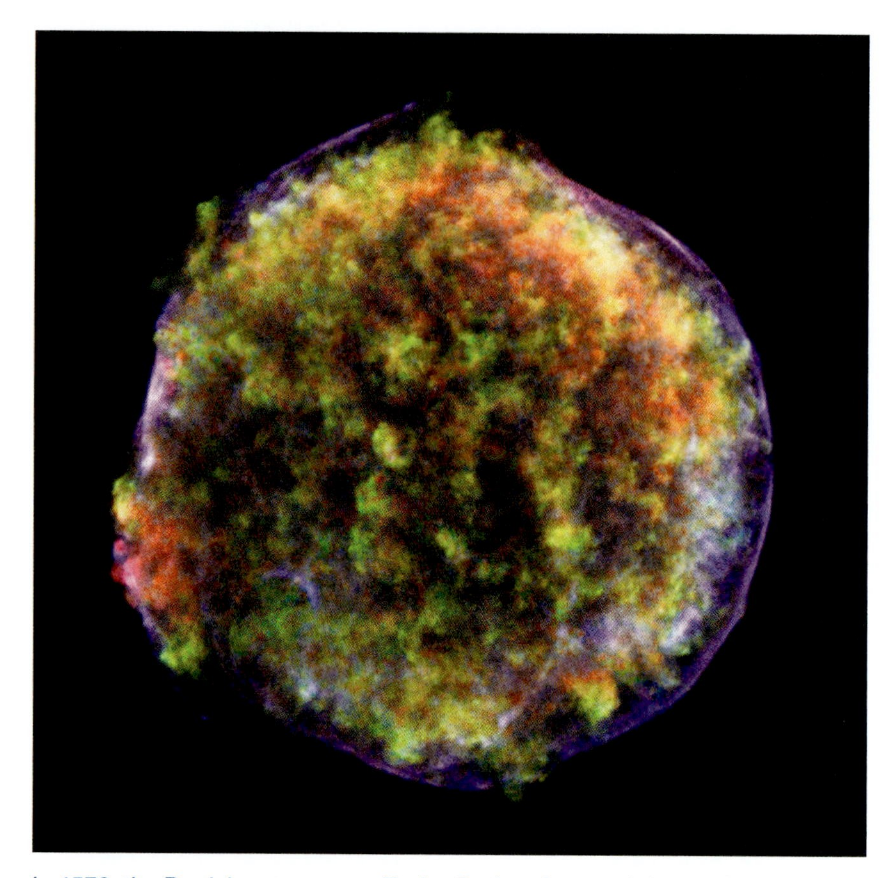

In 1572, the Danish astronomer Tycho Brahe observed the explosion of a massive star—an event that became known as Tycho's supernova. This image of Tycho's supernova remnant was captured in 2003 by NASA's *Chandra X-ray Observatory (CXO).* It shows an expanding bubble of multimillion-degree debris (green and red) inside a more rapidly moving shell of extremely high electrons (filamentary blue). *(NASA/CXC/Rutgers/J. Warren, J. Hughes, et al.)*

neutrons that accompanies a supernova explosion is responsible for the rapid neutron capture process (or *r-process*). It is these rapid neutron capture reactions that form the radioactive nuclei of the heaviest elements found in nature, such as thorium-232 and uranium-238. The violently explosive force of the supernova also hurls stellar-minted elements throughout interstellar space in a dramatic shower of stardust.

The expelled stardust eventually combined with interstellar gas. This elementally enriched interstellar gas then became available to create a new generation of stars and, for many of these next generation stars, a family of companion planets. About 4.6 billion years ago, humans' solar system (including planet Earth) formed from just such an elementally enriched giant cloud of hydrogen and helium gas. All things animate and inanimate on Earth are the natural by-products of these ancient astrophysical processes. The chemical elements that enrich planet Earth and support life came from the stars. As the American astronomer Carl Sagan (1934–96) was fond of saying, "We are made of stardust!"

THE MYSTERY OF DARK MATTER AND DARK ENERGY

No discussion of matter is complete without a brief mention of *dark matter* and *dark energy*. However, a more detailed investigation of the mysteries of dark matter and dark energy must remain the focus of other books.

Scientists define dark matter as matter in the universe that cannot be observed directly because it emits very little or no electromagnetic radiation and experiences little or no directly measurable interaction with ordinary matter. Dark matter is also called nonbaryonic matter. This means that dark matter does not consist of baryons, as does ordinary matter. So what is dark matter? Today, no one really knows for certain. While not readily observable in personal experiences here on Earth, dark matter nevertheless exerts a very significant, large-scale gravitational influence within the observable universe.

Starting in the 1930s, astronomers began suspecting the presence of dark matter (originally called *missing mass*) because of the observed velocities and motions of individual galaxies in clusters of galaxies. A galaxy cluster is an accumulation of galaxies (from 10 to hundreds or even a few thousand members) that lie within a few million light-years of each other and are bound by gravitation. Without the presence of this hypothesized dark matter, the galaxies in a cluster would have drifted apart and escaped from each other's gravitational attraction long ago. Scientists are now engaged in a variety of interesting research efforts to better define, understand, and quantify the elusive, gravitationally influential phenomenon they refer to as dark matter.

Dark energy is the generic name now being given by astrophysicists and cosmologists to an unknown cosmic force field hypothesized to be responsible for the recently observed acceleration in the rate of expansion of the universe. The American astronomer Edwin P. Hubble (1889–1953) first proposed the concept of an expanding universe in 1929. He suggested that observations of Doppler-shifted wavelengths of the light from distant galaxies indicated that these galaxies were receding from Earth at speeds proportional to their distances.

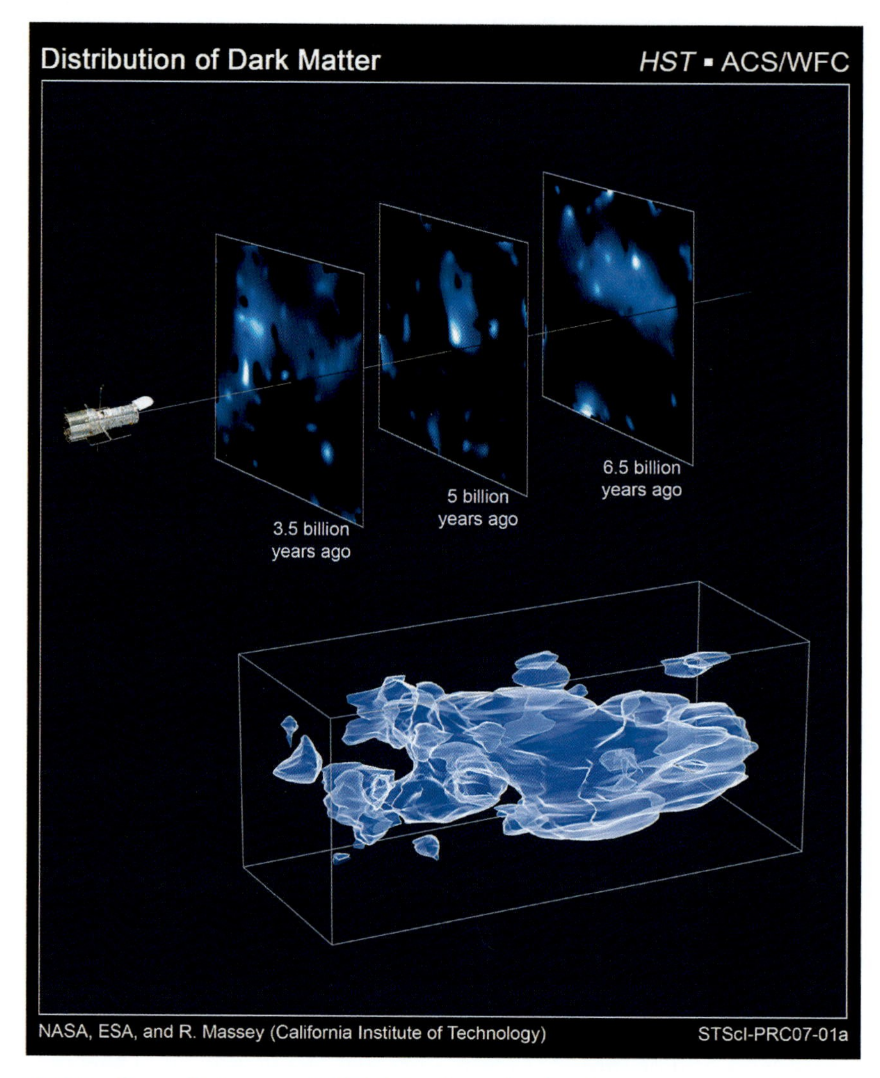

Hubble Space Telescope (HST) data allowed scientists to map the three-dimensional distribution of dark matter in the universe. They used a method called weak gravitational lensing to observe how dark matter deflected the light of faraway galaxies. The top (sliced) image shows how dark matter evolved from 6.5 billion to 3.5 billion years ago. The bottom image shows dark matter clumping as it collapses under gravity. *(NASA, ESA, CalTech)*

In the late 1990s, while scientists were making systematic surveys of very distant Type Ia (carbon detonation) supernovas, they observed that instead of slowing down (as might be anticipated if gravity were the only significant force at work in cosmological dynamics), the rate of recession (that is, the redshift) of these very distant objects appeared to actually be increasing. It was almost as if some unknown force were neutralizing or canceling the attraction of gravity.

Such startling observations proved controversial and very inconsistent with the then popular "gravity only" models of an expanding universe within big bang cosmology. Despite fierce initial resistance within the scientific community, these perplexing observations eventually gained acceptance. Today, carefully analyzed and reviewed supernova data indicate that the rate of expansion of the universe is accelerating—a dramatic conclusion that has tossed modern cosmology into great turmoil.

Cosmologists do not yet have an acceptable answer as to what could be causing this apparent accelerated expansion. Some scientists revisited the cosmological constant (symbol Λ). Albert Einstein inserted this concept into his original general relativity theory to make his revolutionary theory of gravity describe a static universe—that is, a nonexpanding one that had neither a beginning nor an end—but after boldly introducing the cosmological constant as representative of some mysterious force associated with empty space capable of balancing or even resisting gravitational attraction, Einstein decided to abandon the idea. Hubble's announcement of an expanding universe provided the intellectual nudge that encouraged Einstein's decision. Afterward, Einstein personally referred to the notion of a cosmological constant as "my greatest failure."

Physicists are now revisiting Einstein's concept and suggesting that there is possibly a *vacuum pressure force* (a recent name for the cosmological constant). This force appears inherently related to empty space but seems to exert its influence only on a very large scale. Consequently, the mysterious force would have been negligible during the very early stages of the universe following the big bang event but would later have manifested itself and served as a major factor in cosmological dynamics. Since such a mysterious force is neither required nor explained by any of the currently known laws of physics, scientists do not yet have a clear physical interpretation of just what such a mysterious ("gravity resisting") force really means.

On March 7, 2008, NASA released the results of a five-year investigation of the oldest light in the universe, the cosmic microwave background

(CMB). Based on a careful evaluation of the CMB data collected by a NASA spacecraft called the *Wilkinson Microwave Anisotropy Probe (WMAP)*, scientists were able to gain incredible insight into the past and present content of the universe. The *WMAP* data revealed that the current

TAKING A REALLY CLOSE LOOK AT MATTER

Contemplating the heart of matter much like Democritus did in ancient Greece, scientists now propose that every material object, such as the pencil shown here, consists of molecules. The molecules, in turn, are made up of elemental atoms. The atoms themselves contain protons and neutrons located in their tiny central cores. (The exception is elemental hydrogen, which has just a single proton serving as its nucleus.) Even so, the modern view of matter does not stop there, because each neutron or proton contains several incredibly tiny pieces of matter called quarks. Particle physicists are now busy studying the nature and behavior of quarks in an effort to unravel the scientific mystery of why matter has mass.

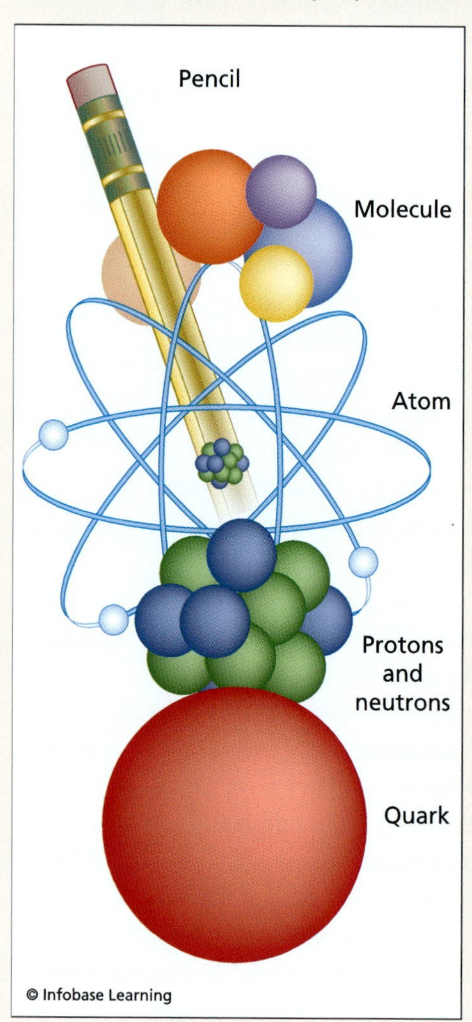

Starting with a common object such as a pencil, modern science suggests this progressively more detailed view of matter. This illustration is obviously not drawn to scale, because a new pencil is about 200 mm long, while an individual quark has an estimated diameter of less than 1 × 10^{-15} mm. One interesting conclusion from this artistic perspective is that so-called solid matter is mostly empty space. *(DOE/FNAL)*

contents of the universe include about 5 percent "ordinary" atoms, the building blocks of stars, planets, and people. Dark matter constitutes 23 percent of the universe. Dark matter is different from the atoms of ordinary matter. It does not emit or absorb light and has only been detected indirectly by its gravitational influence. Finally, about 72 percent of the current universe is composed of dark energy, which acts like a type of matter-repulsive antigravity phenomenon. One of the central challenges facing scientists in this century is how to understand the cosmic roles of ordinary matter, dark matter, and dark energy.

Astrophysical observations, such as the *WMAP* data, suggest that the content of the current universe and the early universe (about 380,000 years after the big bang) are quite different from each other. This difference implies the persistent influence of a cosmic tug of war between energy (radiant and dark) and matter (baryonic and nonbaryonic).

THE STANDARD MODEL

Scientists have developed a quantum-level model of matter they refer to as the *standard model.* This comprehensive model explains (reasonably well) what the material world consists of and how it holds itself together. As shown in the accompanying figure, physicists need only six quarks and six leptons to explain ordinary matter. Despite the hundreds of different particles that have been detected, all known matter particles are actually combinations of quarks and leptons. Furthermore, quarks and leptons interact by exchanging force carrier particles. The most familiar lepton is the electron (e) and the most familiar force carrier particle is the photon.

The standard model is a reasonably good theory and has been verified to excellent precision by numerous experiments. All of the elementary particles that make up the standard model have been observed through experiments, but the standard model does not explain everything of interest to scientists. One obvious omission is the fact that the standard model does not include gravitation.

In the standard model, the six quarks and six leptons are divided into pairs, or generations. The lightest and most stable particles compose the first generation, while the less stable and heavier elementary particles make up the second and third generations. As presently understood by scientists, all stable (ordinary) matter in the universe consists of particles that belong to the first generation. Elementary particles in the second

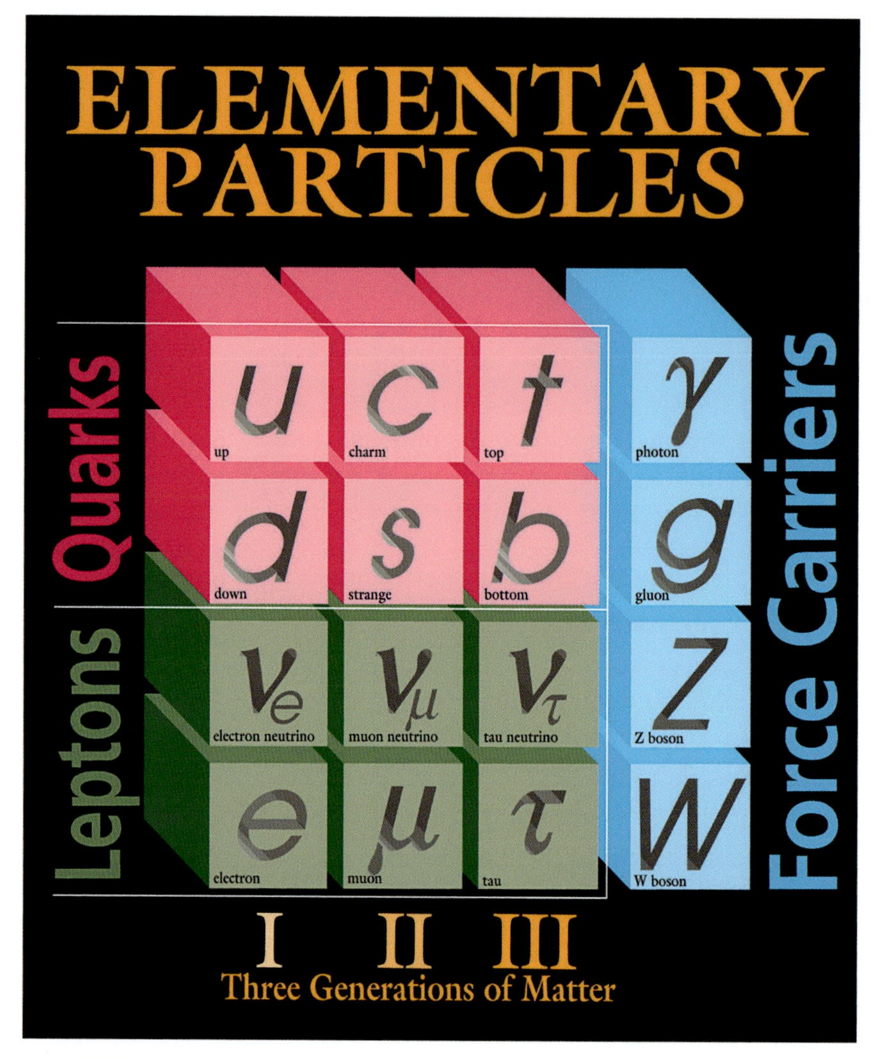

Scientists developed the standard model to explain the complex interplay between elementary particles and force carriers. *(DOE/FNAL)*

and third generation are heavier and have very short lifetimes, decaying quickly to the next most stable generation.

The everyday world of normal human experience involves just three of these building blocks: the up quark, the down quark, and the electron. This simple set of particles is all that nature requires to make protons and neutrons and to form atoms and molecules. The electron neutrino (ν_e) rounds out the first generation of elementary particles. Scientists have observed the electron neutrino in the decay of other particles. There may be other elementary building blocks of matter to explain dark matter, but such building blocks have not yet been observed in experiments.

Elementary particles transmit forces among one another by exchanging force-carrying particles called bosons. The term *boson* is the general name scientists have given to any particle with a spin of an integral number (that is, 0, 1, 2, etc. . . .) of quantum units of angular momentum. Carrier particles of all interactions are bosons. Mesons are also regarded as bosons. The term honors the Indian physicist Satyendra Nath Bose (1894–1974).

As represented in the accompanying figure, the photon (γ) carries the electromagnetic force and transmits light. The gluon *(g)* mediates the strong force and binds quarks together. The *W* and *Z* bosons represent the weak force and facilitate the decay of heavier (more energetic) particles into lower-mass (less energetic) ones. The only fundamental particle predicted by the standard model that has not yet been observed is a hypothetical particle called the Higgs boson. In 1964, the British theoretical physicist Peter Higgs (1929–) hypothesized that this type of particle may explain why certain elementary particles (such as quarks and electrons) have mass and other particles (photons) do not. If found by research scientists this century, the Higgs boson (sometimes called the *God particle*) could play a major role in refining the standard model and shedding additional light on the nature of matter at the quantum level. If nature does not provide scientists with a Higgs boson, then they will need to postulate other forces and particles to explain the origin of mass and to preserve the interactive components of the standard model they have already verified by experiments.

Finally, the force of gravity is not yet included in the standard model. Some physicists suggest that gravitational force may be associated with another hypothetical particle called the *graviton*. Another major challenge facing scientists this century is to develop a quantum formulation of gravitation that encircles and supports the standard model. The harmonious blending of general relativity (which describes gravitation on a cosmic scale) and quantum mechanics (which describes the behavior of matter on the atomic scale) would represent another incredible milestone in humankind's search for the meaning of substance.

CHAPTER 2

Physical Behavior of Matter

In many research activities and technical applications, scientists and engineers find it helpful to discuss matter in terms of certain macroscopic (or bulk) physical properties. Other times, a microscopic (or molecular) perspective provides more meaningful insight. This chapter discusses how both perspectives are useful in describing the nature and behavior of matter.

UNDERSTANDING THE PHYSICAL AND CHEMICAL PROPERTIES OF MATTER

Scientists often compare physical and chemical properties in their efforts to fully characterize substances. A physical property of a substance is a measurable behavior or physical characteristic. Some of the more common physical properties used by scientists are mass, density, temperature, structure, and hardness. Other physical properties include a substance's melting point and boiling point. Heat capacity, thermal conductivity, electric conductivity, solubility, and color are other physical properties.

The chemical properties of a substance describe how that substance reacts with other substances. On a microscopic scale, the atoms of the substance experience change due to their interaction with neighboring atoms. For example, electron trading among neighboring atoms makes or breaks chemical bonds. Changes in ionization state (from electrically neutral to positive or negative) may also take place at the atomic level. As a result of such activities, some substances experience corrosion, others experience combustion (burning) or explosive decomposition, while other substances (such as helium) remain inert and refuse to interact with other materials in their environment. Scientists often use electron configurations to correlate the chemical properties of the elements that appear in the periodic table. (See appendix.)

The physical change of a substance typically involves a change in the physical appearance of that sample of matter, but the substance that undergoes this physical change may not display any change in its chemical composition or identity. Consider adding heat to an uncovered pot of water on a stove. The water eventually starts to boil and then experiences a physical change by becoming a hot gas, commonly called *steam*. Scientists say that water experiences a physical change of state when it transforms to steam. Despite this change in physical state, steam retains the chemical identity of water.

MASS, VOLUME, AND DENSITY

Mass (m) is the amount of material present in an object. This fundamental physical property identifies how much stuff (matter) makes up a particular physical object. Scientists use the kilogram (kg) as the basic unit of mass in a widely used measurement system called the International System of Units (SI).

The American customary system of units employs the pound-mass (lbm) as the basic unit of mass. This unit system traces its heritage back to the system of weights and measures used in the United Kingdom before 1824. The measurement system emerged in a somewhat haphazard fashion over the previous centuries. For example, the basic unit of length, the foot (ft), traces its ancestry back to attempts by 14th-century British monarchs to standardize a unit of length for use throughout their kingdom. They chose a physical reference that had been used since the beginning of human civilization—the length of a man's foot. As a result of a somewhat unscientific averaging and selection process in medieval England, the British foot emerged as being equal to 12 inches, with the inch being approximately the width of a man's thumb.

Astronomical references provided the basic unit of time, the second (s). In addition to length (ft), mass (lbm), and time (s), the American customary system also inherited another basic unit, the pound-force (lbf). The distinction between the pound-mass and the pound-force is often blurred in daily activities. Both "pounds" are physically related by an arbitrary definition, which stated that "a one pound-mass object weighs one pound-force at sea level on Earth." Technical novices and professionals alike frequently forget that this arbitrary "pound" equivalency is valid only at sea level on the surface of Earth. An arbitrary historic arrangement established the pound-force as a *basic* unit in the American customary system. This arbitrary arrangement still causes much confusion, since scientists prefer to treat force as a *derived unit,* based on Isaac Newton's second law of motion. Scientists define a derived unit as a unit of measurement that can be obtained by multiplying or dividing various combinations of the basic units that make up the unit system.

The International System (SI) has nine basic units, from which all other units can be derived. The basic SI units for length, mass, and time are the meter (m) (British spelling: *metre*), the kilogram (kg), and the second (s). These are based on easily reproducible natural standards and international agreement. The other six basic units in the SI system are the ampere (A, as a measure of electric current), the candela (cd, as a measure of luminous intensity), the kelvin (K, as a measure of absolute thermodynamic temperature), the mole (mol, as a measure of the amount of substance based on Avogadro's principle), the radian (rad, for plane angles), and the steradian (sr, for solid angles). There are numerous supplementary and derived SI units, such as the becquerel (Bq, as a measure of radioactivity) and the newton (N, as a measure of force). Scientists define the newton as the amount of force that gives a mass of one kilogram an acceleration of one meter per second-squared. One major advantage is that SI units use the number 10 as a base. Multiples or submultiples of SI units are reached by multiplying or dividing by 10. For example, one kilometer (km) represents a length of 1,000 meters, while a nanometer (nm) is an extremely small length, just 1×10^{-9} m.

SI measurements play an important role in science and technology throughout the world. Unfortunately, many technical and most commercial activities in the United States still involve use of American customary system units. The only other two countries where the SI system is not used are Liberia (in western Africa) and the Myanmar Republic (formerly Burma).

Volume (V) is a measure of the three-dimensional space occupied by a solid object or by a given mass of fluid, such as a contained liquid or a confined gas. Scientists and engineers use simple mathematical formulas to

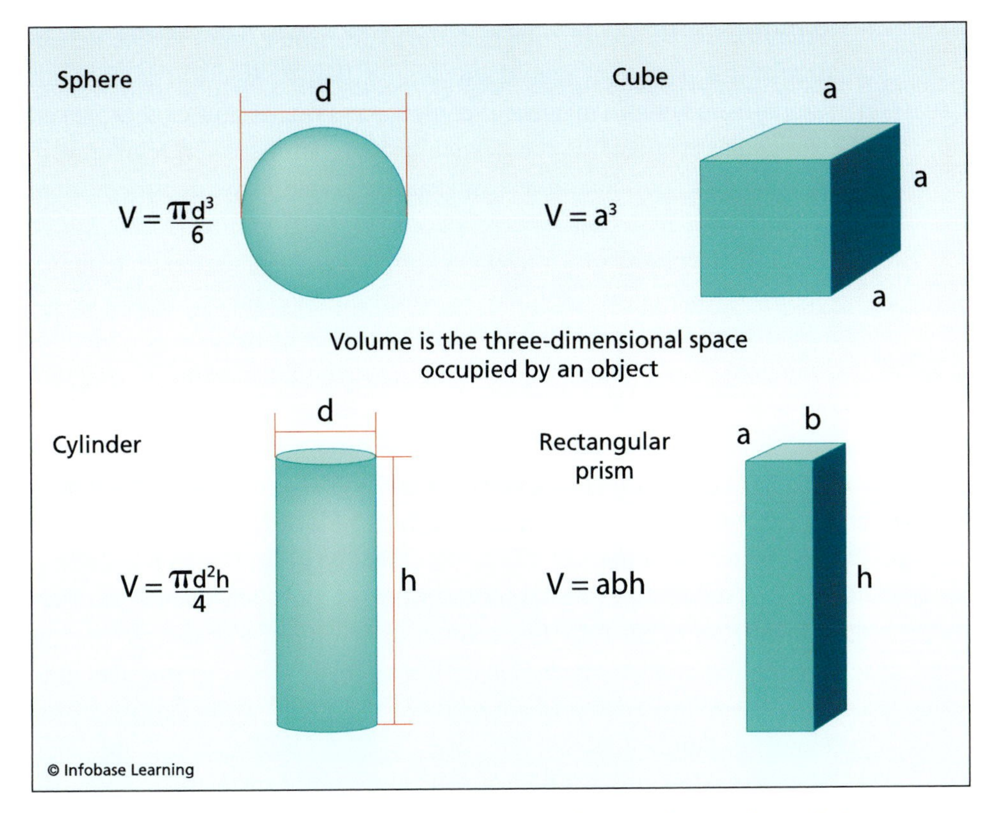

The mathematical formulas for calculating the volumes of several regularly shaped solid objects *(NASA)*

calculate the volume of objects that have well-defined linear (straight-line) features or circular shapes. They use integral calculus to determine the volume of more complex curved shapes. The technique involves approximating the curved-shaped object by a collection of many tiny cubes and concentric cylindrical shells and then summing the volumes of all the tiny cubes and cylinders. They use liquid displacement techniques (based on Archimedes' principle) to estimate the volume of irregularly shaped solid objects.

To assist in more easily identifying and characterizing different materials, scientists devised the material property called *density,* one of the most useful macroscopic physical properties of matter. Solid matter is generally denser than liquid matter and liquid matter denser than gases. Scientists define density as the amount of mass contained in a given volume. They frequently use the lower-case Greek letter rho (ρ) as the symbol for density in technical publications and equations.

Scientists use the density of a material to determine how massive a given volume of that particular material would be. Density is a function of both the atoms from which a material is composed as well as how closely packed the atoms are arranged in the particular material. At room temperature (nominally 68°F [20°C]) and one atmosphere pressure, the density of some interesting solid materials is as follows: gold, 1,205 lbm/ft^3 (19,300 kg/m^3 [19.3 g/cm^3]); iron, 493 lbm/ft^3 (7,900 kg/m^3 [7.9 g/cm^3]); diamond (carbon), 219 lbm/ft^3 (3,500 kg/m^3 [3.5 g/cm^3]); aluminum, 169 lbm/ft^3 (2,700 kg/m^3 [2.7 g/cm^3]); and bone, 112 lbm/ft^3 (1,800 kg/m^3 [1.8 g/cm^3]).

One cubic meter (35.31 ft^3) of any solid or liquid material commonly found on Earth is a large amount of matter—often too much for just one person to handle easily in a laboratory environment. Imagine trying to move a 35.31 ft^3 (1 m^3) chunk of ice (at 32°F [0°C]), which has a mass of about 2,028 lbm (920 kg); or a 35.31 ft^3 (1 m^3) block of pure gold (at 68°F [20°C]), which has a mass of 42,557 lbm (19,300 kg). In research activities, scientists often use smaller quantities of mass and then employ another (equivalent) set of SI units to express the density of a material. Staying within the SI system, they say that ice has a density of 0.92 grams per cubic centimeter (g/cm^3), and gold a density of 19.3 g/cm^3 at the temperatures previously mentioned.

The element osmium (Os) is a platinum-family metal with a density of 1,411 lbm/ft^3 (22,600 kg/m^3 [22.6 g/cm^3]) at room temperature. This material has the highest density of any element found naturally on Earth. Osmium is a hard, brittle, lustrous blue-white metal with an atomic number (Z) of 76 and an atomic mass (A) of 190. The metal enjoys the highest density title because its atoms are very massive and are packed very closely together in a hexagonal crystalline lattice. For comparison, the metal mercury (Hg), which is a liquid at room temperature, has a density of 849 lbm/ft^3 (13,600 kg/m^3 [13.6 g/cm^3]).

Like most gases at room temperature (nominally 68°F [20°C]) and one atmosphere pressure, oxygen (O) has a density of just 0.083 lbm/ft^3 (1.33 kg/m^3 [1.33×10^{-3} g/cm^3])—a value that is about 1,000 times lower than the density of most solid or liquid materials normally encountered on Earth's surface.

It is important to recognize that the physical properties of matter are often interrelated. Namely, when one physical property, such as temperature, changes (increases or decreases), other physical properties, such as volume or density, also change. Over the past three centuries, scientists have learned how to define the behavior of materials by developing special mathematical expressions called *equations of state.* They developed these mathematical relationships using both theory and empirical data from many carefully conducted laboratory experiments.

PRESSURE

Scientists define *pressure* (P) as force per unit area. The most commonly encountered unit of pressure in the American customary system is pounds-force per square inch (psi). In the SI unit system, the fundamental unit of pressure is called the pascal (Pa) in honor of the French scientist Blaise Pascal (1623–62), who conducted many pioneering experiments in fluid mechanics. One pascal represents a force of one newton (N) exerted over an area of one square meter—that is, 1 Pa = 1 N/m^2. One psi is approximately equal to 6,895 Pa.

Anyone who has plunged into the deep end of a large swimming pool and then descended to the bottom of the pool has personally experienced the phenomenon of hydrostatic pressure. Hydrostatic pressure is the pressure at a given depth below the surface of a static (nonmoving) fluid. As Pascal observed in the 17th century, the greater the depth, the greater the pressure.

Atmospheric pressure plays an important role in many scientific and engineering disciplines. Scientists use the following values in an effort to standardize their research activities: at sea level, one atmosphere (1 atm) = 760 mm Hg (exactly) = 29.92 in (Hg) = 14.695 psi = 1.01325×10^5 Pa.

One important feature of Earth's atmosphere is that the density in a column of air above a point on the planet's surface is not constant. Rather, density, along with atmospheric pressure, decreases with increasing altitude until it becomes essentially negligible in outer space. The pioneering work of Pascal and the Italian physicist Evangelista Torricelli (1608–47) guided other scientists in measuring and characterizing Earth's atmosphere.

Scientists and engineers often treat rigid, solid bodies as incompressible objects. Generally, very high values of force are needed to compress or deform a rigid, solid body. (The important solid mechanics concepts of deformation, stress, and strain appear in chapter 4.) Unlike rigid solids, fluids are materials that can flow, so engineers use pressure differentials to move fluids. They design pumps to move liquids (often treated as incompressible fluids), while they design fans to move gases (compressible fluids). An incompressible fluid is assumed to have a constant value of density; a compressible fluid has a variable density. One of the interesting characteristics of gases is that, unlike solids or liquids, they can be compressed into smaller and smaller volumes.

TEMPERATURE

While temperature is one of the more familiar physical variables, it is also one of the most difficult to quantify. Scientists suggest that on the macroscopic scale, temperature is the physical quantity that indicates how hot or cold an object is relative to an agreed upon standard value. Temperature defines the natural direction in which energy will flow as heat—namely, from a higher temperature (hot) region to a lower temperature (cold) region. Taking a microscopic perspective, temperature indicates the speed at which the atoms and molecules of a substance are moving.

Scientists recognize that every object has this physical property called temperature. They further understand that when two bodies are in thermal equilibrium, their temperatures are equal. A thermometer is an instrument that measures temperatures relative to some reference value. As part of the scientific revolution, creative individuals began using a variety of physical principles, natural references, and scales in their attempts to quantify the property of temperature.

In about 1592, Galileo Galilei attempted to measure temperature with a device he called the thermoscope. Although Galileo's work represented the first serious attempt to harness the notion of temperature as a useful scientific property, his thermoscope, while innovative, did not supply scientifically significant temperature data.

The German physicist Daniel Gabriel Fahrenheit (1686–1736) was the first person to develop a thermometer capable of making accurate, reproducible measurements of temperature. In 1709, he observed that alcohol expanded when heated and constructed the first closed-bulb glass thermometer with alcohol as the temperature-sensitive working fluid. Five years later (in 1714), he used mercury as the thermometer's working fluid. Fahrenheit selected an interesting three-point temperature reference scale for his original thermometers. His zero point (0°F) was the lowest temperature he could achieve with a chilling mixture of ice, water, and ammonium chloride (NH_4Cl). Fahrenheit then used a mixture of just water and ice as his second reference temperature (32°F). Finally, he chose his own body temperature (recorded as 96°F) as the scale's third reference temperature.

After his death, other scientists revised and refined the original Fahrenheit temperature scale, making sure there were 180 degrees between the freezing point of water (32°F) and the boiling point of water (212°F) at one atmosphere pressure. On this refined scale, the average temperature of the human body appeared as 98.6°F. Although the Fahrenheit temperature scale is still used in the United States, most of the other nations in the world have adopted another relative temperature scale, the Celsius scale.

In 1742, the Swedish astronomer Anders Celsius (1701–44) introduced the relative temperature scale that now carries his name. He initially selected the upper (100-degree) reference temperature on his new scale as the freezing point of water and the lower (0-degree) reference temperature as the boiling of water at one atmosphere pressure. He then divided the scale into 100 units. After Celsius's death in 1744, the Swedish botanist and zoologist Carl Linnaeus (1707–78) introduced the present-day Celsius scale thermometer by reversing the reference temperatures. The modern Celsius temperature scale is a relative temperature scale in which the range between two reference points (the freezing point of water at 0°C and the boiling point of water at 100°C) are conveniently divided into 100 equal units, or degrees.

Scientists define a *relative temperature scale* as one that measures how far above or below a certain temperature measurement is with respect to a specific reference point (such as the triple point of water). The individual degrees, or units, in a relative scale are determined by dividing the relative scale between two known reference temperature points (such as the freezing and boiling points of water at one atmosphere pressure) into a convenient number of temperature units (such as 100 or 180).

Despite considerable progress in thermometry in the 18th century, scientists still needed a more comprehensive temperature scale—namely, one that included the concept of absolute zero, the lowest possible temperature, at which molecular motion ceases. The Irish-born British physicist William Thomson (1824–1907), first baron Kelvin of Largs, proposed an absolute temperature scale in 1848. Kelvin's temperature scale was quickly embraced by the scientific community. The proper SI unit term for temperature is *kelvin* (without the word *degree*), and the proper symbol is K (without the symbol °).

The Fahrenheit scale and Celsius scale are relative temperature scales. Scientists established the two absolute temperature scales—the Rankine [R] scale and Kelvin [K] scale—by setting the lowest temperature of the scales at absolute zero. Absolute temperature values are always positive, but relative temperatures can have positive or negative values. Scientists generally use absolute temperatures in such disciplines as physics, astronomy, and chemistry; engineers use either relative or absolute temperatures

WHY ICE FLOATS

Most solids and liquids expand as their temperatures increase. Water (H_2O) is a very interesting exception. When the temperature of pure (fresh) water is raised from 32°F (0°C) to 39.2°F (4°C), the liquid actually contracts (shrinks) slightly. However, above 39.2°F (4°C), water expands (just like other liquids) as the temperature increases. This means that the density of pure water passes through a maximum value at 39.2°F (4°C) temperature. At all other temperatures, the density of water is less than the maximum value of 62.4 lbm/ft^3 (1 g/cm^3). Ice floats because it has a lower density, about 57.1 lbm/ft^3 (0.915 g/cm^3) at 32°F (0°C). One consequence is that freshwater lakes and ponds freeze from the top down, allowing fish and other aquatic life to survive in slightly warmer (say, 39.2°F [4°C]), deeper parts even after the surface of the lake or pond has frozen over.

The answer to water's interesting physical behavior lies in molecular bonding. Liquid water consists of molecules that are packed more closely together due to covalent bonding; ice consists of a crystalline molecular structure in which the individual hydrogen and oxygen atoms in each molecule are pushed a bit farther apart, resulting in a lower value of density.

A leopard seal resting on a newly calved chunk of glacial ice *(NOAA)*

The formation of sea ice (versus newly calved, freshwater, glacial ice) is a bit more complicated and depends on such additional factors as salinity levels and sea surface environmental conditions. Scientists have observed that the Arctic and Antarctic Oceans consist of layers of water with different properties, a condition that also influences the sea ice formation process.

in thermodynamics, heat transfer analyses, and mechanics, depending upon the nature of the problem.

When the atoms in a solid heat up, they vibrate more and tend to push the neighboring atoms a little farther away. For most materials, this atomic scale behavior then manifests itself as a slight increase in the dimensions as they are heated. The phenomenon, called *thermal expansion,* provided scientists a measurable link between temperature and other macroscopic properties of a solid substance and the substance's atomic structure.

Consider a crystalline substance in which the individual atoms are assumed to be held together by an orderly arrangement of stiff springs, representing chemical bonds. (Chemical bonds are discussed later in this chapter.) As the temperature increases, the atoms vibrate more, and the substance's size increases slightly to accommodate the increased atomic motions. Laboratory experiments and engineering experience indicate that most solids and liquids expand as their temperatures increase. One very notable exception is water.

Scientists created a figure of merit, called the *coefficient of linear expansion* (symbol α), to help them quantify thermal expansion phenomena and to allow them to compare the expansion characteristics of various solid materials. Consider a metal (say, copper) rod of initial length L. If scientists raise the rod's temperature by an amount ΔT, the increase in the rod's length ΔL due to thermal expansion is given by the formula: $\Delta L = \alpha L \Delta T$. At room temperature conditions, the linear expansion coefficient for copper is 9.4×10^{-6} per °F (17×10^{-6} per °C). Other important solid substances have different linear expansion coefficients. Lead has an α value of 16.1×10^{-6} per °F (29×10^{-6} per °C); aluminum has α equal to 12.8×10^{-6} per °F (23×10^{-6} per °C); concrete, 6.7×10^{-6} per °F (12×10^{-6} per °C); steel, 6.1×10^{-6} per °F (11×10^{-6} per °C); and diamond, 0.67×10^{-6} per °F (1.2×10^{-6} per °C).

CHEMICAL ELEMENTS FOUND ON EARTH

The chemical properties of matter remained a mystery until woven into an elegant intellectual tapestry called the periodic table. This section briefly describes the significance of the periodic table, which appears in the appendix of this book.

In 1869, the Russian chemist Dmitri Ivanovich Mendeleev published *The Principles of Chemistry,* a textbook within which he introduced the periodic law of chemical elements. This law stated that the properties of the elements are a periodic function of their atomic weights. Other chemists had tried in vain to bring some type of order to the growing number of elements. Mendeleev's attempt was the most successful arrangement up to that time and soon was embraced by chemists around the world. The modern version of the periodic table was greatly expanded in content and meaning by the rise of modern atomic theory and quantum mechanics in the first three decades of the 20th century.

The modern periodic table evolved from Mendeleev's pioneering work. However, several major scientific developments—such as the discovery of radioactivity, the discovery of the electron, the introduction of the nuclear atom model, and the emergence of quantum mechanics—were necessary before the periodic table could assume its current structure and detailed information content. Today, scientists continue to explore the nature of matter as they attempt to create new superheavy elements.

As of July 2010, scientists had observed and identified 118 elements. Of these, 94 occur naturally on Earth. However, six of the 94 elements that occur on Earth exist naturally only in trace quantities. These wispy elements are technetium (Tc, atomic number 43), promethium (Pm, atomic number 61), astatine (At, atomic number 85), francium (Fr, atomic number 87), neptunium (Np, atomic number 93), and plutonium (Pu, atomic number 94). The advent of nuclear accelerators and reactors has made synthetic production of these scarce elements possible. Consequently, many, such as neptunium and plutonium, are now available in significant quantities. The modern nuclear age also made available countless relatively short-lived radioisotopes of all the elements.

Chemists and physicists now correlate the properties of the elements portrayed in the periodic table with their electron configurations. Since in a neutral atom the number of electrons equals the number of protons, scientists find it convenient to arrange the elements in the periodic table in order of their increasing atomic number (Z). The modern periodic table has seven horizontal rows (called periods) and 18 vertical columns (called groups). The properties of the elements in a particular row vary across it, thereby providing the concept of periodicity. Scientists refer to the elements contained in a particular vertical column as a group, or family. (See appendix.)

There are several versions of the periodic table used in modern science. The International Union of Pure and Applied Chemistry (IUPAC) recommends labeling the vertical columns from 1 to 18, starting with hydrogen (H) as the top of group 1 and ending with helium (He) as the top of group number 18. The IUPAC further recommends labeling the periods (rows) from 1 to 7. Hydrogen (H) and helium (He) are the only two elements found in period (row) number 1. Period number 7 starts with francium (Fr) and includes the actinide series as well as the transactinides (very short–lived, human-made, superheavy elements).

The row (or period) in which an element appears in the periodic table tells scientists how many electron shells an atom of that particular element possesses. The column (group number) lets scientists known how many electrons to expect in an element's outermost electron shell. Scientists call an electron residing in an atom's outermost shell a *valence electron.* Chemists have learned that it is these valence electrons that determine the chemistry of a particular element. The periodic table is structured such that all the elements in the same column (group) have the same number of valence electrons. Consequently, the elements that appear in a particular column (group) display similar chemistry.

Chemists and physicists have found it useful to describe families of elements using such terms as *alkali metals, alkaline earth metals, transition metals, nonmetals,* and *noble gases.* A few of these groups (or families) of elements are summarized briefly here. For a more comprehensive discussion of the modern periodic table, the reader is referred to any contemporary college-level chemistry textbook.

The *alkali metals* are the family of elements found in group 1 (column 1) of the periodic table. They include lithium (Li), sodium (Na), potassium (K), rubidium (Rb), cesium (Cs), and francium (Fr). These metals react vigorously with water to produce hydrogen gas. The *alkali earth metals* are the family of elements found in group 2 (column 2) of the periodic table. They include beryllium (Be), magnesium (Mg), calcium (Ca), strontium (Sr), barium (Ba), and radium (Ra). These elements are chemically reactive, but less so than the alkali metals. The *noble gases* are the inert gaseous elements found in group 18 (column 18) of the periodic table. These elements include helium (He), neon (Ne), argon (Ar), krypton (Kr), xenon (Xe), and radon (Rn). The noble gases do not readily enter into chemical combinations with other elements. Human-produced nuclear reactions result in the production of certain radioactive isotopes of krypton and xenon, while radon is a naturally occurring source of radioactivity.

The two long rows that appear below the periodic table are horizontal expansions that correspond to the position of lanthanum (La) or actinium (Ac). The lanthanoid (formerly lanthanide) series contains 15 elements that start with lanthanum (La, atomic number 57) and continue through lutetium (Lu, atomic number 71). All the elements in the lanthanoid series closely resemble the element lanthanum. Scientists often collectively refer to the 15 elements in the lanthanoid series along with the elements scandium (Sc) and yttrium (Y) as the rare earth elements or the rare earth metals. The actinoid (formerly actinide) series of heavy metallic elements starts with element 89 (actinium) and continues through element 103 (lawrencium). These elements are all radioactive and together occupy one position in the periodic table. The actinoid series includes uranium (atomic number 92) and all the human-made transuranium elements.

A transuranium (or transuranic) element is an element with an atomic number (Z) greater than that of uranium. These elements extend from neptunium (Np), with an atomic number of 93, to lawrencium (Lr), with an atomic number of 103. All transuranium elements are essentially human-made and radioactive, although extremely small trace amounts of natural neptunium and plutonium have been detected on Earth. Some of the more significant transuranium elements are neptunium (93), plutonium (94), americium (95), curium (96), berkelium (97), and californium (98).

Scientists sometimes include in their discussion of transuranium elements the collection of human-made superheavy elements that starts with element number 104 (called rutherfordium [Rf]) and currently extends to element 118. These superheavy elements are also called *transactinide elements* because they appear in row (period) 7 of the periodic table, after the element actinium (Ac). Fleeting amounts (one to several atoms) of all these elements, have been created in complex laboratory experiments that involve the bombardment of special transuranium target materials with high-velocity nuclei.

CHEMICAL BONDS

Chemical energy is liberated or absorbed during a chemical reaction. In such a reaction, energy losses or gains usually involve only the outermost electrons of the atoms or ions of the system undergoing change. Here, a chemical bond of some type is established or broken without disrupting the original atomic or ionic identities of the constituents.

Scientists discovered that the atoms in many solids are arranged in a regular, repetitive fashion and that the atoms in this structured array are held together by interatomic forces called *chemical bonds*. Chemical bonds play a significant role in determining the properties of a substance. There are several different types of chemical bonds. These include ionic bonds, covalent bonds, metallic bonds, and hydrogen bonds.

An *ionic bond* is created by the electrostatic attraction between ions that have opposite electric charges. Sodium chloride (NaCl), better known as table salt, is a common example of ionic bonding. Sodium (Na) is a silvery metal that has one valence electron to lose and form the cation (positive ion) Na^+. Chlorine (Cl) is a pale, yellow-green gas that has seven valence electrons and readily accepts another electron to form the anion (negative ion) Cl^-. The bond formed in creating a sodium chloride molecule is simply electrostatic attraction between sodium and chlorine. When a large number of NaCl molecules gather together, they form an ionic solid that has a regular crystalline structure. In summary, the ionic bond discussed here is the result of the sodium atom transferring a valence electron to the valence shell of a chlorine atom, forming the ions Na^+ and Cl^- in the process. Every Na^+ ion within the salt (NaCl) crystal is surrounded by six Cl^- ions, and every Cl^- ion by six Na^+ ions.

The second type of chemical bond is called the *covalent bond.* In the covalent bond, two atoms share outer-shell (valence) electrons. The molecular linkage takes place because the shared electrons are attracted to the positively charged nuclei (cores) of both atoms. One example is the

A close-up view of sodium chloride (NaCl) crystals in a water bubble within a 50-mm diameter metal loop photographed by a crew member doing research in the Destiny laboratory on the *International Space Station* on March 13, 2003 *(NASA)*

hydrogen molecule (H_2), in which two hydrogen atoms are held tightly together. In 1916, the American chemist Gilbert Newton Lewis (1875–1946) proposed that the attractive force between two atoms in a molecule was the result of electron-pair bonding, now called covalent bonding. During the following decade, developments in quantum mechanics allowed other scientists to quantitatively explain covalent bonding. In most molecules, the atoms are linked by covalent bonds. Generally, in most molecules hydrogen forms one covalent bond; oxygen, two covalent bonds; nitrogen, three covalent bonds; and carbon, four covalent bonds. In the water molecule, the oxygen atom bonds with two hydrogen atoms. The H—O bonds form as the oxygen atom shares two pairs of electrons, while each hydrogen atom shares only one pair of electrons, which scientists refer to as *polar covalent bonds.*

In *metallic bonding,* electrons are freely distributed so that many metal atoms share them. Immersed in this sea of negative electrons, the positive metal ions form a regular crystalline structure. Consider the metal sodium as an example. This metal is made up of individual sodium atoms each of which has 11 electrons. In metallic bonding, every sodium atom releases one valence electron, which then moves throughout the metallic crystal attracted to the positive Na^+ ions. It is this attraction that holds the metallic crystal together.

Finally, *hydrogen bonding* is a weak to moderate attractive force due to polarization phenomena. This type of bonding occurs when a hydrogen atom that is covalently bonded in one molecule is at the same time attracted to a nonmetal atom in a neighboring atom. Hydrogen bonding typically involves small atoms characterized by high electronegativity, such as oxygen, nitrogen, and fluorine. Chemists define *electronegativity* as the attraction of an atom in a compound for a pair of shared electrons in a chemical bond. Under certain conditions during physical state transitions, hydrogen bonding occurs in the water molecule. Hydrogen bonds are also common in most biological substances.

CHAPTER 3

The Gravity of Matter

How does matter give rise to gravity? As part of the scientific revolution, Galileo Galilei and Sir Isaac Newton provided the first useful quantitative interpretation of the relationship between matter and gravity. Albert Einstein extended their work when he introduced general relativity early in the 20th century. For Einstein, gravity represented the warping of the space-time continuum by massive objects. This chapter presents both the classical and modern physics interpretations of one of nature's most familiar yet mysterious phenomena—gravity.

THE NOTION OF GRAVITY

Physicists currently recognize the existence of four fundamental forces in nature: gravity, electromagnetism, the strong force, and the weak force. Gravity and electromagnetism are part of everyday experience. These two forces have an infinite range, which means they exert their influences over very great distances. The other two forces are much less familiar, since both act only within the realm of the atomic nucleus, but unfamiliarity does not imply that these two short-range forces are insignificant. Rather, the strong and weak forces govern the fascinating mass-energy conversions that light up individual stars and make entire galaxies visible.

Gravity is the attractive force that tugs on people and holds them on the surface of Earth. Gravity also keeps the planets in orbit around the Sun and causes the formation of stars, planets, and galaxies.

Where there is matter, there is gravity. In a simple definition, gravity is the pull material objects exert on each other. While mastering the art of walking under the watchful eyes of their parents, toddlers often instinctively respond to the pervasive tug of Earth's gravity by gently tumbling to the ground. The oil and vinegar in a bottle of salad dressing naturally separates (differentiates) as the bottle rests motionless on a shelf or table. A baseball thrown up into the air will rise to a certain height, pause at the apex of its trajectory, and then begin its downward trip to Earth under gravity's influence. These are all commonly encountered examples of nature's most relentless, long-range force.

Early humans experienced gravity and sometimes took advantage of its all-pervading presence in their constant struggle for survival. When they harvested ripe fruit that fell from tall trees or pit-trapped large game animals, prehistoric peoples used gravity without understanding what this mysterious natural phenomenon was.

Despite great scientific progress since the scientific revolution, gravity remains one of nature's most mysterious phenomena. Galileo Galilei is often remembered as the first astronomer to use a telescope to view the heavens and conduct early astronomical observations that helped inflame the scientific revolution. However, he was also the physicist who founded the science of mechanics and provided Newton the underlying data and ideas that allowed the British physicist to develop the laws of motion and the universal law of gravitation.

Galileo Galilei was born in Pisa on February 15, 1564 (scientists commonly refer to Galileo by his first name only). In 1585, he left the university without receiving a degree and focused his activities on the physics of solid bodies. The motion of falling objects and projectiles intrigued him. In 1589, Galileo became a mathematics professor at the University of Pisa. He was a brilliant lecturer, and students came from all over Europe to attend his classes. This circumstance quickly angered many senior but less capable faculty members. To make matters worse, Galileo often used his tenacity, sharp wit, and biting sarcasm to win philosophical arguments at the university. His tenacious and argumentative personality earned him the nickname "The Wrangler."

In the late 16th century, European professors usually taught natural philosophy (physics) as metaphysics, an extension of Aristotelian philosophy. Before Galileo's pioneering contributions, physics was not regarded as an observational or experimental science, but through his skillful use of mathematics and innovative experiments, he changed that approach and established the important approach now known as the *scientific method*. Galileo's research activities constantly challenged the 2,000-year tradition of ancient Greek learning.

Aristotle had stated that heavy objects would fall faster than lighter objects. Galileo disagreed and held the opposite view that, except for air resistance, the two objects would fall at the same rate regardless of their masses. It is not certain whether he personally performed the legendary musket ball versus cannonball drop experiment from the Leaning Tower in Pisa to prove this point. He did, however, conduct a sufficient number of experiments with objects rolling or sliding down inclined planes to upset Aristotelian thinking and create the science of mechanics.

According to popular scientific legend, Galileo dropped a cannonball and a musket ball from the Leaning Tower of Pisa to demonstrate that, save for wind resistance, objects fall at the same rate under the influence of gravity. While Galileo never documented this particular experiment in his notes, he did perform a large number of experiments with objects rolling or sliding down inclined planes. Through these important experiments, Galileo upset Aristotelian "physics" and created the modern science of mechanics. *(U.S. Air Force)*

During his lifetime, Galileo was limited in his motion experiments by an inability to accurately measure small increments of time. No one had yet developed a timekeeping device capable of accurately measuring tenths, hundredths, or thousandths of a second. Despite this severe impediment, he conducted many important experiments that produced remarkable insights into the physics of free fall and projectile motion. Newton would build upon Galileo's pioneering work to create

1.37 second for the geologist's hammer and falcon's feather to simultaneously hit the surface. Remember, the local acceleration of gravity on the surface of the Moon is just 5.25 feet/s^2 (1.6 m/s^2), because the Moon is much smaller and less massive than Earth. Scott's simple demonstration validated the universality of the physical laws that emerged from the great intellectual accomplishments of Galileo, Newton, and other scientists who looked up at the Moon, planets, and stars and wondered. One of the most important contributions of Western civilization to the human race is the development of modern science and the emergence of the scientific method. During the intellectually turbulent period of the 17th century, men of great genius such as Galileo developed important physical laws and experimental techniques that helped human beings explain the operation and behavior of the physical universe.

NEWTON'S MECHANICAL UNIVERSE

Sir Isaac Newton was the brilliant though introverted British physicist, mathematician, and astronomer whose law of gravitation, three laws of motion, development of calculus, and design of a new type of reflecting telescope made him one of the greatest scientific minds in human history. Through the patient encouragement and financial support of the British mathematician Sir Edmund Halley (1656–1742), Newton published his great work, *The Principia* (full title: *Mathematical Principles of Natural Philosophy*), in 1687. This monumental book transformed the practice of physical science and completed the scientific revolution started by Nicholas Copernicus (1473–1543) a century earlier. Newton's three laws of motion and universal law of gravitation still serve as the basis of classical mechanics.

Newton was born prematurely in Woolsthorpe, Lincolnshire, on December 25, 1642 (using the former Julian calendar). His father had died before Newton's birth, and this event contributed to a very unhappy childhood. In 1665, he graduated with a bachelor's degree from Cambridge University without any particular honors or distinction.

Following graduation, Newton returned to the family farm to avoid the plague, which had broken out in London. For the next two years, he pondered mathematics and physics at home, and this self-imposed exile laid the foundation for his brilliant contributions to science. By Newton's own account, one day on the farm he saw an apple fall to the ground and began to wonder if the same force that pulled on the apple also kept the

Moon in its place. At the time, heliocentric cosmology as expressed in the works of Copernicus, Galileo, and Kepler was becoming more widely accepted (except where banned on political or religious grounds), but the mechanism for planetary motion around the Sun remained unexplained.

By 1667, the plague epidemic had subsided, and Newton returned to Cambridge as a minor fellow at Trinity College. The following year, he received his master of arts degree and became a senior fellow. In about 1668, he constructed the first working reflecting telescope, an important new astronomical instrument. This telescope design earned Newton a great deal of professional acclaim, including eventual membership in the Royal Society.

In 1669, Isaac Barrow, Newton's former mathematics professor, resigned his position so that the young Newton could succeed him as Lucasian Professor of Mathematics. This position provided Newton the time to collect his notes and properly publish his work—a task he was always tardy to perform.

Shortly after his election to the Royal Society in 1671, Newton published his first scientific paper. While an undergraduate, Newton had used a prism to refract a beam of white light into its primary colors (red, orange, yellow, green, blue, and violet.) Newton reported this important discovery to the Royal Society. To Newton's surprise, this pioneering work was immediately attacked by Robert Hooke (1635–1703), an influential member of the society.

This attack was the first in a lifelong series of bitter disputes between Hooke and Newton. Generally, Newton only skirmished lightly then quietly retreated. This was Newton's lifelong pattern of avoiding direct conflict. When he became famous later in his life, Newton would start a controversy, withdraw, and then secretly manipulate others, who would then carry the brunt of the battle against Newton's adversary. Newton's famous conflict with the German mathematician Gottfried Leibniz (1646–1716) over credit for the invention of calculus followed such a pattern. Through Newton's clever manipulation, the calculus controversy even took on nationalistic proportions as carefully coached pro-Newton British mathematicians bitterly argued against Leibniz and his supporting group of German mathematicians.

In August 1684, Halley traveled to Newton's home at Woolsthorpe. During this visit, Halley convinced the reclusive genius to address the following puzzle about planetary motion: What type of curve does a planet describe in its orbit around the Sun, assuming an inverse square law of black holes by speculating about their theoretical properties and then looking for perturbations in the observable universe that provide telltale signs that a massive, invisible object matching such theoretical properties is possibly causing these perturbations. Here on Earth, a pattern of ripples on the surface of an otherwise quiet but murky pond could indicate that a large fish is swimming just below the surface. Similarly, astrophysicists look for detectable ripples in the observable portions of the universe to support theoretical predictions about the behavior of black holes.

The very concept of a mysterious black hole exerts a strong pull on both scientific and popular imaginations. Data from space-based astronomical observatories, such as NASA's *Chandra X-ray Observatory,* have moved black holes from the purely theoretical realm to a dominant position in observational astrophysics. Strong evidence is accumulating that black holes not only exist, but that very large ones called *supermassive black holes* may contain millions or billions of solar masses. Astrophysicists further suggest that the supermassive black hole may function like a carnivorous cosmic monster lurking at the center of every large galaxy.

How did the idea of a black hole originate? The first person to publish a paper about black holes was John Michell (1724–93), a British geologist, amateur astronomer, and clergyman. In his 1784 scientific paper, Michell suggested the possibility that light (then erroneously believed to consist of tiny particles of matter subject to influence by Newton's law of gravitation) might not be able to escape from a sufficiently massive star. Michell was a competent astronomer who successfully investigated binary-star system populations. He further suggested that although no "particles of light" could escape from such a massive object, astronomers might still infer its existence by observing the gravitational influence the massive object would exert on nearby celestial objects.

The French mathematician and astronomer Pierre-Simon, marquis de Laplace (1749–1827), introduced a similar concept in the late 1790s when he likewise applied Newton's law of gravitation to a celestial body so massive that the force of gravity would prevent any light particles from escaping. Neither Michell nor Laplace used the term *black hole* to describe their postulated very massive heavenly bodies. In fact, the term did not enter the lexicon of astrophysics until the American astrophysicist John Archibald Wheeler (1911–2008) introduced it in 1967. Both of these 18th-century black hole speculations were on the right track but suffered from incomplete and inadequate physics.

The needed breakthrough in physics took place a little more than a century later, when Einstein introduced relativity theory. He replaced the Newtonian concept of gravity as a force between two or more masses with the novel concept that gravity was associated with the way massive objects distorted the fabric of space-time. As suggested by Einstein in the mid-1910s, the more massive an object, the greater is its ability to distort the local space-time continuum.

Shortly after Einstein introduced general relativity, the German astronomer Karl Schwarzschild (1873–1916) discovered that Einstein's relativity equations led to the postulated existence of a very dense object into which other objects could fall but out of which no objects could ever escape. In 1916, Schwarzschild wrote the fundamental equations that describe a black hole. He also calculated the size of the *event horizon,* or boundary of no return, for this incredibly dense and massive celestial object. The dimension of a black hole's event horizon now bears the name *Schwarzschild radius* in his honor.

The notion of an event horizon implies that no information about events taking place inside this distance can ever reach outside observers. The event horizon is not a physical surface; it represents the start of a special region of the universe that is disconnected from normal space and time. Although scientists cannot see beyond the event horizon into a black hole, they believe that time stops there.

Inside the event horizon, the escape speed exceeds the speed of light. Outside the event horizon, escape is possible. It is important to remember that the event horizon is not a material surface. It is the mathematical definition of the point of no return, the point where all communication is lost with the outside world. Inside the event horizon, the laws of physics as humans currently understand them do not apply. Once anything crosses this boundary, it will disappear into an infinitesimally small point known as a singularity. Scientists cannot observe, measure, or test a singularity, since it, too, is a mathematical definition.

The little that scientists presently know about black holes comes from looking at the effects they have on the surrounding region of space *outside* the event horizon. As more powerful astronomical observatories study the universe across all portions of the electromagnetic spectrum this century, scientists will be able to construct continually better theoretical models of the black hole. For now, physicists must remain content to regard the black hole as gravity's ultimate triumph over matter.

The more massive the black hole, the greater the extent of its event horizon. For example, the event horizon for a stellar black hole containing five solar masses is just 9.3 mi (15 km) from the singularity. In the Milky Way galaxy, a supernova occurs on average once or twice every 100 years. Since the galaxy is about 13 billion years old, scientists suggest that about 200 million supernovas have occurred, creating neutron stars and black holes. Astronomers have identified at least a dozen stellar black hole candidates (containing between four and 20 or more solar masses) by observing how certain stars appear to wobble under the suspected gravitational influence of nearby massive but invisible companions.

X-ray binary systems offer another way to search for candidate black holes. Observations of iron atoms in the hot gases orbiting several stellar black hole candidates have allowed scientists to investigate the gravitational effects and spin of these suspected black holes. Data collected by NASA's *Chandra X-ray Observatory* and the European Space Agency's *XMM-Newton* spacecraft suggest that the gravity of a spinning black hole shifts X-ray signals from the (accretion cloud) iron atoms to lower energies, producing the strongly skewed X-ray signals.

The orbit of a particle near a black hole (but outside the event horizon) depends on the curvature of space around the black hole, which also depends on how fast the black hole is spinning. A spinning black hole drags space around with it and allows atoms to orbit nearer to the black hole than is possible for a nonspinning black hole. The tighter orbit means stronger gravitational effects, which in turn means more of the X-rays from iron atoms are shifted to lower energies. The most detailed studies of stellar black holes to date indicate that not all black holes spin at the same rate.

Once matter crosses the event horizon and falls into a black hole, only three physical properties appear to remain relevant: total mass, net electric charge, and total angular momentum. Recognizing all that black holes must have mass, physicists have proposed four basic stellar black hole models. The Schwarzschild black hole (first postulated in 1916) is a static, that is, nonspinning, black hole that has no charge and no angular momentum. The Reissner-Nordström black hole (introduced in 1918) has an electric charge but no angular momentum—that is, it is not spinning. In 1963, the New Zealand mathematician Roy Patrick Kerr (1934–) applied general relativity to describe the properties of a rapidly rotating but uncharged black hole. Astrophysicists think that this model is the most likely "real-world" black hole because the massive stars that formed

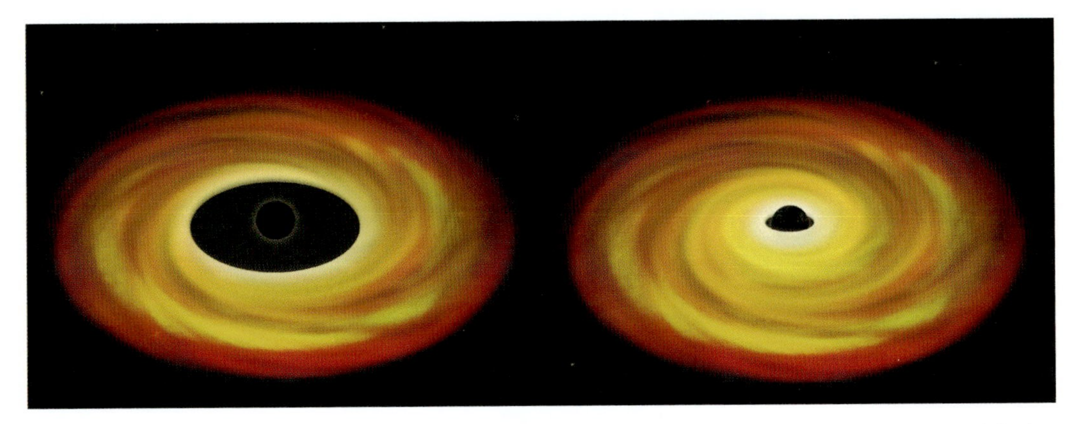

This artist's rendering shows a nonrotating black hole (on the left) and a rotating black hole (on the right). The most detailed studies of stellar black holes to date indicate that not all black holes spin at the same rate. *(NASA/CXC/M. Weiss)*

them would have been rotating. The final stellar black hole model has both charge and angular momentum. Called the Kerr-Newman black hole, this theoretical model appeared in 1965. Since most astrophysicists currently think that rotating black holes are unlikely to have a significant electric charge, the uncharged spinning black hole remains the leading theoretical candidate.

In comparison to stellar black holes, the event horizon for a supermassive black hole consisting of 100 million solar masses is about 186 million miles (300 million km) from its singularity—twice the distance of Earth to the Sun. Astronomers suspect that such massive objects exist at the centers of all large galaxies. No one knows for sure how such supermassive black holes formed. One hypothesis is that over billions of years, relatively small stellar black holes formed by supernovas began devouring neighboring stars in the star-rich centers of large galaxies and eventually became supermassive black holes.

CHAPTER 4

Fundamentals of Materials Science

Materials science is the very broad technical field that encompasses the study and application of all materials. This chapter examines some of the fundamental principles that govern the practice of materials science.

BASIC CONCEPTS

Persons engaged in the practice of materials science seek to understand the formation, structure, and properties of materials on scales that range from the microscopic (atomic) to the macroscopic (that is, as manifested in daily human experiences). Fundamental to the study of materials is the establishment of quantitative and predictable relationships that describe the way a material is processed (produced) and performs (behaves) under certain conditions.

Materials scientists seek to understand the important relationship between how a particular material's atoms are arranged and its characteristic physical and chemical properties. They first divide materials into two general forms: solids and fluids. They then subdivide solid materials into two basic categories: crystalline and amorphous (or noncrystalline). This categorization is based upon how the atoms or molecules of each type of solid material are internally arranged. Scientists and engineers regard metals (such as copper, gold, steel, and lead), ceramics (such as aluminum oxide [Al_2O_3] and magnesium oxide [MgO]), and semiconductor materials (such as silicon arsenide [SiAs] and gallium arsenide [GaAs]) as crystalline solids because their atoms form an ordered internal structure.

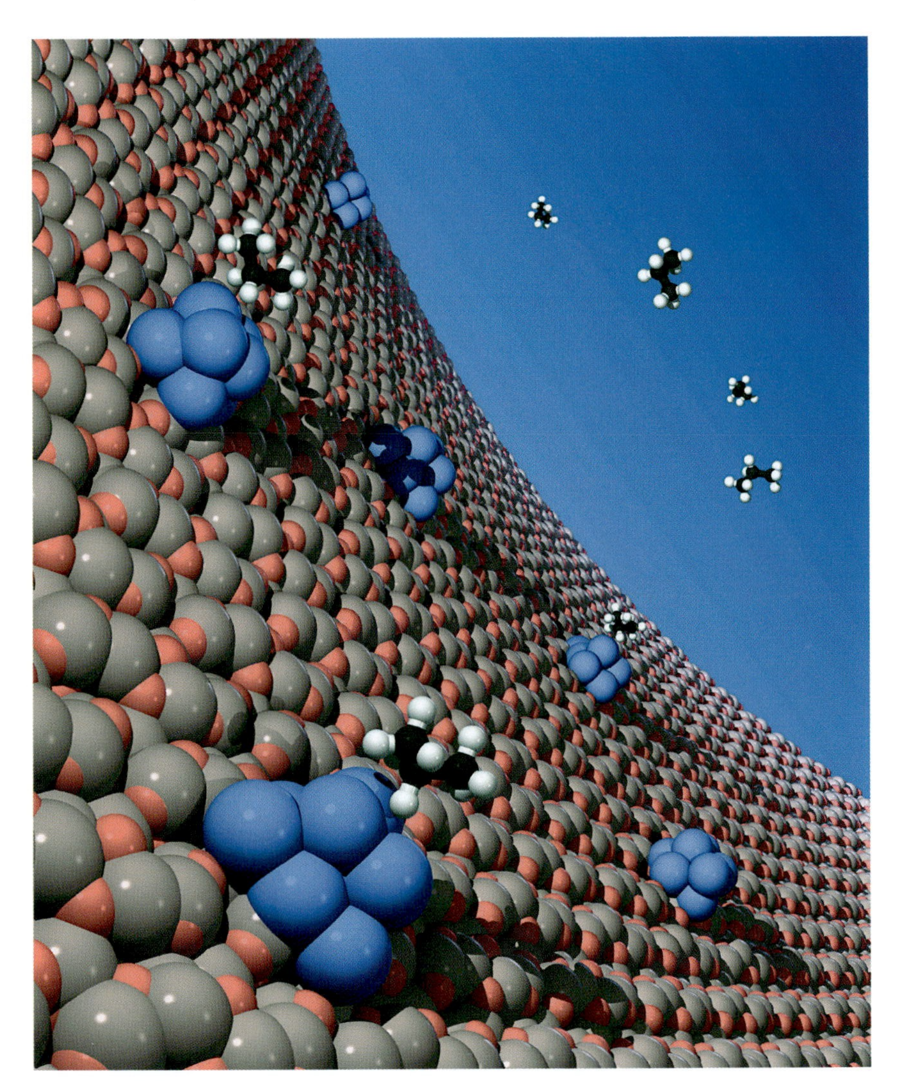

The artists' rendering shows clusters of 8 to 10 platinum atoms (blue atoms) deposited in pores of a crystalline aluminum oxide (Al_2O_3) (gray and pink atoms) membrane. The platinum atoms are highly active and selective catalysts for the oxidative dehydrogenation of propane (C_3H_8) molecules (white and black atoms). *(DOE/ANL; artists, M. Sternberg and F. Mehmood)*

The majority of polymers (such as plastics) and glasses are examples of amorphous solids. The term *amorphous* means that the solid substance does not have any long-range atomic or molecular arrangement. Obsidian is a naturally occurring amorphous solid that is formed as a glassy, extrusive igneous rock due to volcanic action. Prehistoric peoples favored obsidian to make their knives, arrowheads, spearheads, and tools.

The shape of a solid is an important physical characteristic. On a macroscopic scale, a solid's shape may establish its ability to rest in a stable manner on a surface or else to fit properly in some desired configuration. On the microscopic (or atomic) scale, the orderly placement of atoms and molecules establishes the structure and characteristics of crystalline solids. Materials scientists use their mathematical knowledge of two-dimensional shapes (called polygons) and three-dimensional shapes (called polyhedrals) to describe and understand the basic crystalline structures characteristic of many solid materials. Some well-known polygons are the triangle (three sides), the quadrilateral (four sides), the pentagon (five sides), and the hexagon (six sides). Mathematicians define regular polygons as those for which all sides are the same length and all angles are identical. The equilateral triangle and the square are familiar examples of regular polygons.

Certain three-dimensional shapes, called the five regular polyhedrals or the five Platonic solids after the ancient Greek philosopher, are often encountered in science. These are the tetrahedron (formed by four triangles), the cube (formed by six squares), the octahedron (formed by eight triangles), the dodecahedron (formed by 12 pentagons), and the icosahedron (formed by 20 triangles). In addition, spheres, right circular cylinders, and right circular cones are solid shapes often formed or closely approximated in nature. An orange, a tree trunk, and a cone volcano are respective examples.

Scientists and engineers define a material's *strength* as its ability to resist deformation or changes in shape. The strength of a solid is directly linked to the strength of the chemical bonds that hold the substance's atoms and molecules together. When characterizing a material's strength, scientists recognize that there are three basic types of strength that must be considered: tensile strength, compressive strength, and shear strength. Tensile strength is the material's ability to resist being pulled apart. Compressive strength is the ability of a material to resist crushing or being squeezed. Finally, shear strength is a material's ability to resist torsion (being twisted) or shearing (one molecular plane sliding over another).

MATERIALS PROCESSING IN SPACE

The advent of the space age in 1957 opened up the research potential of the microgravity environment found in an Earth-orbiting spacecraft. The availability of extended periods of microgravity (often abbreviated as μg)—as, for example, onboard the operational *International Space Station (ISS)*—promises to bring new and unique opportunities for the science of materials processing.

During a 1983 mission of NASA's *Challenger,* an astronaut materials scientist heated mercury iodide source material in a specially designed microgravity vapor crystal growth system furnace. The vaporized source material then traveled to a seed crystal and condensed, forming a much larger crystal suitable for use in sensitive nuclear radiation detection instruments. *(NASA)*

One important objective of microgravity materials science is to gain an improved understanding of how gravity-driven phenomena impact the process of solidification and the crystal growth of materials. On Earth, buoyancy-driven convection, hydrostatic pressure, and sedimentation can create defects in the internal structure of materials, which irregularities in turn alter their properties. Certain crystals, such as mercury-iodide (HgI), are sufficiently dense and delicate that the influence of their own weight subjects them to strain during growth. Such strain-induced deformations degrade the susceptible crystals' overall performance. Materials scientists are attempting to use microgravity to develop heat-treated, melted, and resolidified crystals and alloys that are significantly free of such deformations.

Over the next few years, space-based materials processing research will emphasize both scientific and commercial goals. Potential space-manufactured products include specialized crystals, metals, ceramics, glasses, and biological materials. As research in these areas progresses, customized new materials and manufactured products could become available this century for use in space as well as on Earth.

Engineers often compare the strength of various materials using a graphical presentation of relevant material property data. The stiffness of a solid material is one of the most important properties considered in engineering design work. Engineers use Young's modulus of elasticity (discussed shortly) as a measure of a solid material's stiffness, or ability to resist tension, compression, or torsion. The accompanying illustration depicts the relative stiffnesses of entire classes of materials (such as metals, polymers, foams, and ceramics) versus the corresponding range of density values, another important design parameter. As a typical starting point in these comparisons, the solid materials are assumed uniform in composition and isotropic in elastic behavior. In materials science, the word *isotropic* means that the substance has identical physical properties in all directions. Sometimes the assumption of uniformity in all directions is a good approximation of natural behavior; other times this assumption leads to major inaccuracies. Scientists perform many careful measurements and controlled experiments on reference samples of candidate substances to establish whether certain physical properties are isotropic or anisotropic.

As shown in the figure, metals are typically very strong (stiff) and have high densities, while foams are generally quite flexible and have low densities. Metals and alloys constitute an important category of engineered materials. Engineered materials include structural materials, composite materials, electrical conductors, and magnetic materials. The properties of metals and alloys are closely linked to their crystalline structures and chemical properties. The mechanical strength and corrosion resistance of an alloy are determined by the internal arrangement of atoms that develops as the alloy solidifies from its molten state.

Polymers are basically very large molecules (sometimes called macromolecules) consisting of numerous small repeating molecular units (called monomers) that are usually joined together by covalent chemical bonds. Natural polymers include silk, rubber, and wool; human-made or manufactured polymers include nylon, polyester, and plastic. How the individual monomers in a polymer are bonded together strongly influences the substance's physical properties. Natural and synthetic rubbers are examples of *elastomers*—amorphous polymers that exhibit well-defined elastic behavior (elasticity). The word *plastic,* when used as a noun, is the popular name that people give to a large family of organic substances (polymers) that can be injected into molds and cast into various shapes or else extruded and drawn into thin filaments and sheets.

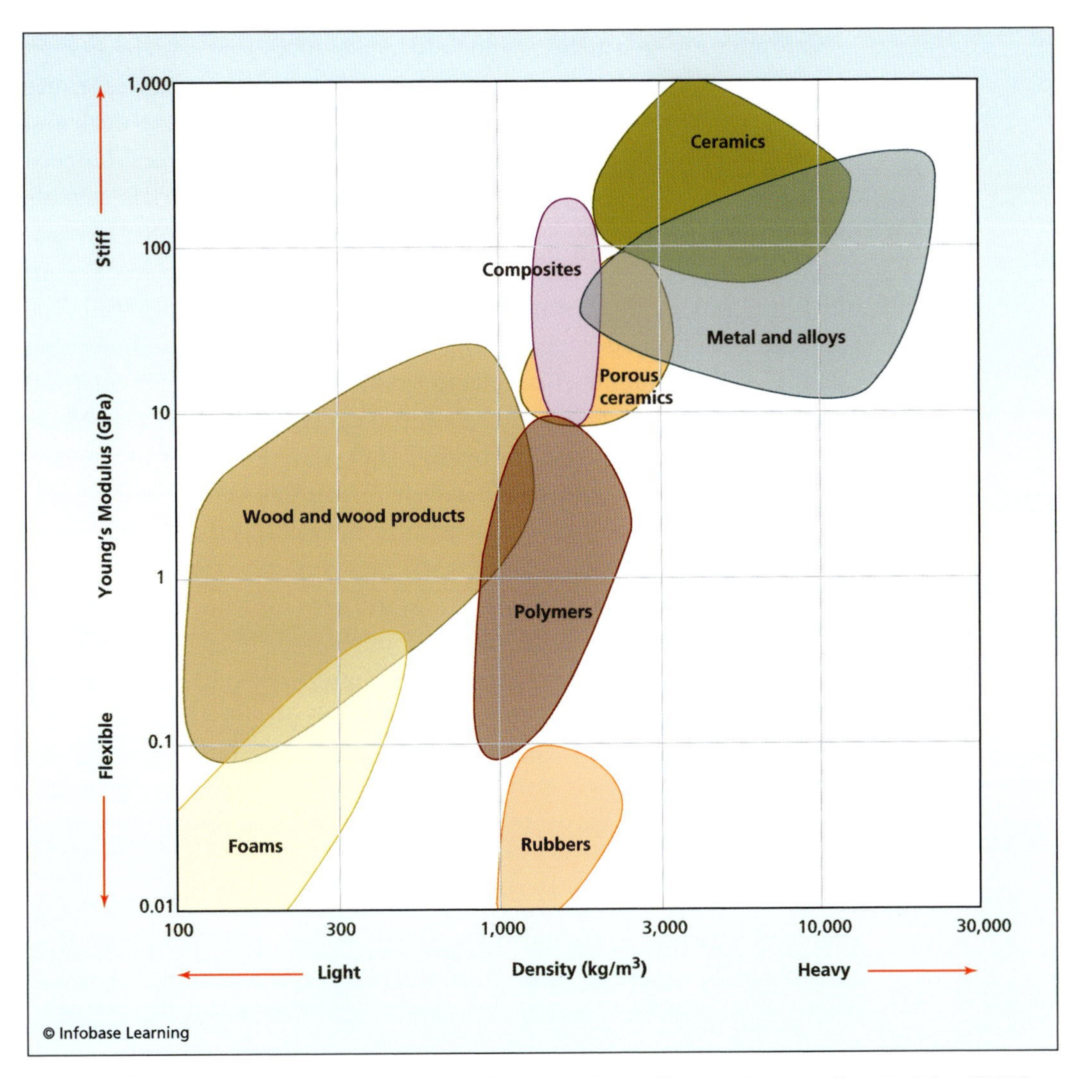

Chart of Young's modulus as a function of density for different classes of materials *(DOE)*

Scientists and engineers have learned how to apply heat and pressure to various organic compounds to create plastic materials that perform a wide range of jobs. Plastics now represent an integral material component of civilization. There are two general types of plastics: thermoplastic and thermosetting. Materials scientists can use heat and pressure to repeatedly soften and remold thermoplastic materials. In contrast, once molded and set in some initial configuration or product, thermosetting materials cannot be reheated and remolded into another useful shape or object.

Ceramics are inorganic nonmetallic substances that usually exhibit high levels of strength at elevated temperatures. The use of ceramics extends back to prehistoric times. Early peoples learned how to work and heat clay to create the first useful ceramics—pottery and bricks. Modern materials scientists have developed advanced ceramic materials suitable for many specialized applications, including as thermal protection systems for aerospace vehicles. The great majority of ceramic materials can resist heat and chemical attack but do not conduct electricity well. The porcelain used in modern toilet bowls and other bathroom facilities is an example of a commonly encountered sanitary application of specialized ceramic materials. Materials such as boron carbide (B_4C), silicon carbide (SiC), and alumina (aluminum oxide [Al_2O_3]) are examples of modern ceramic materials sometimes collectively referred to as technical ceramics. One major difficulty with most ceramic materials is that they are brittle and hard. When these materials fail, they fail catastrophically, often breaking in an irreparable manner. Just ask anyone who has dropped a fancy dinner plate on a hard floor. Upon impact, the once beautiful piece of ceramic dinnerware usually fragments into many odd-shaped pieces that fly off in all directions. An empty metal pot dropped from the same height onto the same floor will bounce and perhaps dent but remain intact and useful.

ELASTICITY

Engineers measure the strength of materials by the capacity of substances, such as metals, concrete, wood, glass, and plastics, to withstand stress and strain. As part of the scientific revolution, scientists began investigating the strengths of materials in an organized, quantitative way.

The British scientist Robert Hooke studied the action of springs in 1678 and reported that the extension (or compression) of an elastic material (such as a spring) takes place in direct proportion to the force exerted on the material. Today, physicists and engineers use *Hooke's law* to quantify the displacement associated with the restoring force of an ideal spring.

Scientists define *elasticity* as the ability of a body to resist a stress (distorting force) and then return to its original shape when the stress is removed. There are several types of stresses: tension (due to a pulling force), compression (due to a pushing force), torsion (due to a rotating or twisting force), and shear (due to internal slipping or sliding). All solid objects are elastic to some degree when they experience deformation. The

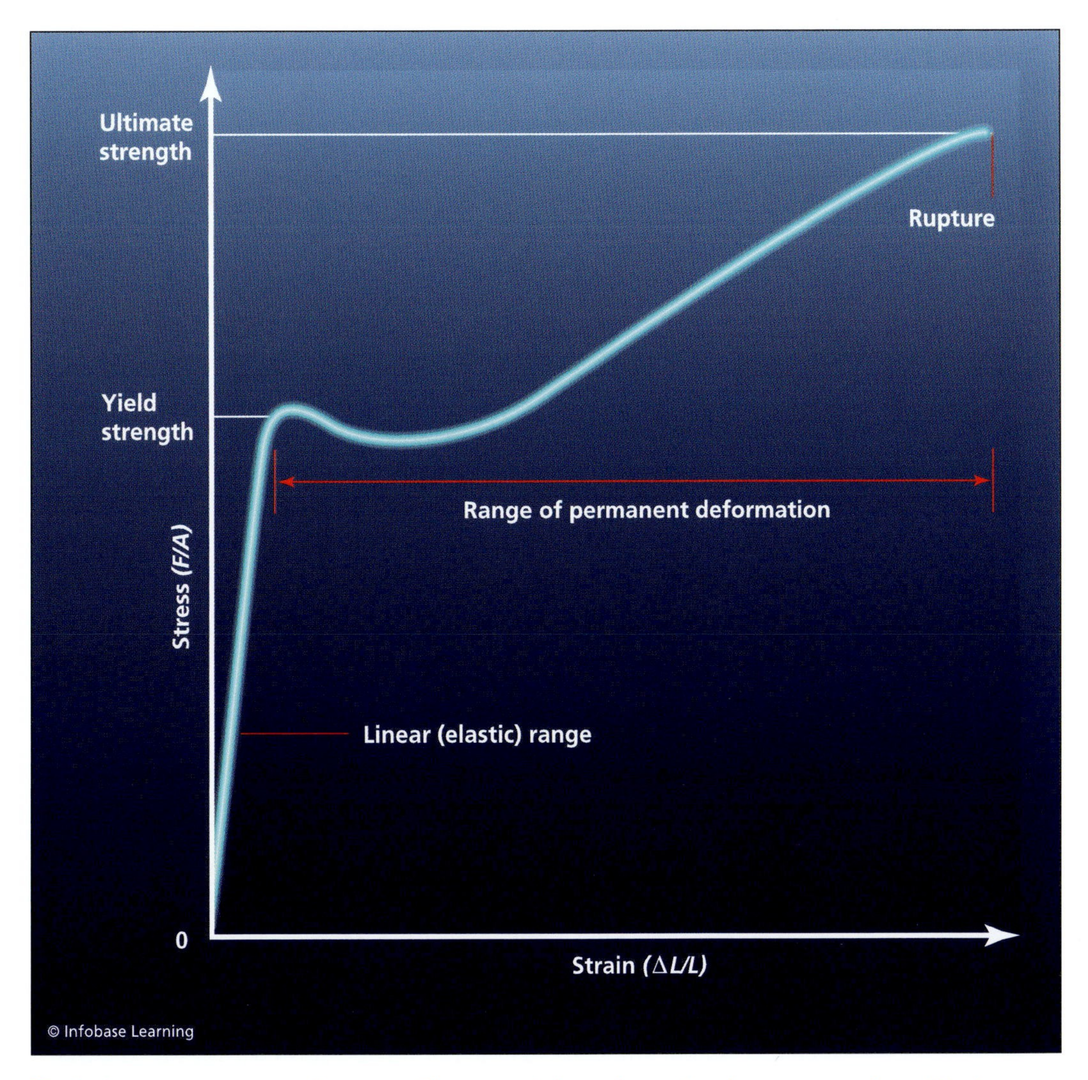

Typical stress versus strain curve. The material specimen (such as a type of steel) being tested deforms permanently when the stress equals the material's yield strength. Additional stress causes the material to further deform and eventually rupture when the stress reaches the material's ultimate strength. *(Author)*

degree of elasticity is a function of the material. According to Hooke's law, within the elastic limit of a particular substance, the stress is proportional to the strain.

If a person pulls on a metal spring (that is, exerts a tension or pulling force), the spring will experience a strain and stretch a certain amount in proportion to the applied tensile force. Within the material's elastic limit, once the person stops pulling on the spring, it restores itself to the original position. (This also occurs with springs that work in compression.) However, if the person keeps pulling on the spring and stretches (deforms) it beyond the material's elastic limit, the spring experiences a plastic deformation. Once the tension force is released after such a plastic deformation, the material, here a metal spring, can no longer return to its original shape and dimensions. If the person keeps pulling on the metal spring after plastic deformation, the metal stretches to the yield point, at which it breaks (ruptures).

The field of solid mechanics deals with the behavior of solid materials under a variety of external influences, including stress forces. Some materials, such as steel, are much better at resisting tensile (pulling) forces. Other materials, such as concrete, are much better at resisting compressive (pushing) forces. The strength of various materials is ultimately determined by the arrangement of the atoms and molecules that make up the material. In the early 19th century, building upon the work of previous scientists such as Hooke and the Swiss mathematician Leonhard Euler (1707–83), the British physician and physicist Thomas Young (1773–1829) proposed the use of a coefficient of elasticity to compare the relative strengths of various materials. Young's modulus (also called the modulus of elasticity) describes the ratio of stress to strain and applies to many materials over their respective elastic ranges—that is, where Hooke's law is valid. At the yield point, a material begins to experience permanent deformation, departing Hooke's law behavior and entering plastic behavior. The continued application of stress during plastic deformation eventually causes the material to exceed its ultimate strength and rupture or fail.

When a person stretches a rubber band and stays within the material's elastic limit, the rubber band returns to its original dimensions when the tension (pulling) forces cease. However, if a person stretches the rubber band too much, it quickly ruptures or snaps apart. Unlike the rubber band, a bar of taffy is not very elastic but has an interesting range of plastic behavior. When a person begins to pull on a piece of taffy from both ends, it starts stretching. Eventually, it stretches into ("necks-down" to) a very long and narrow piece that ultimately breaks apart.

Engineers and scientists define *stress* as the force (tensile or compressive) applied per unit area on an object. Typical units to describe stress are pounds-force per square inch (psi) and newtons per square meter (N/m^2). Engineers recognize that such units are the same as pressure, which in the metric system is usually expressed as pascals (Pa). *Strain* is a measure of an object's deformation (change in size) due to stress. Strain is a dimensionless quantity, so Young's modulus, which describes the ratio of stress to strain, has the units of pressure and is usually expressed in psi or pascals.

Examination of the elastic properties of some typical engineering materials should prove informative at this point. Structural steel has a density of 491 lbm/ft^3 (7.86 g/cm^3), Young's modulus of elasticity of 2.9 $\times 10^7$ psi (200×10^9 Pa), yield strength of 3.6×10^4 psi (250×10^6 Pa), and ultimate strength of 5.8×10^4 psi (400×10^6 Pa). Aluminum has a density of 169 lbm/ft^3 (2.71 g/cm^3), Young's modulus of elasticity of 1.02×10^7 psi (70×10^9 Pa), a yield strength of 1.38×10^4 psi (95×10^6 Pa), and an ultimate strength of 1.60×10^4 psi (110×10^6 Pa). Concrete has a density of 145 lbm/ft^3 (2.32 g/cm^3), Young's modulus of 4.35×10^6 psi (30×10^9 Pa), and an ultimate strength of 5,800 psi (40×10^6 Pa) (in compression). Polystyrene (plastic) has a density of 65.6 lbm/ft^3 (1.05 g/cm^3), Young's modulus of 43,500 psi (3×10^9 Pa), and an ultimate strength of 6,960 psi (48×10^6 Pa). For many materials, Young's modulus is approximately the same for both tension and compression, but certain materials are stronger in compression than tension and vice versa. Concrete is very weak in tension but strong in compression, so engineers usually express its ultimate strength for compression only.

Steel and other ductile metals generally exhibit an interesting behavior when stress forces continue beyond the material's yield strength. Scientists use the term *ductility* to characterize a material's ability to plastically deform while under (tensile) stress before reaching its ultimate strength and rupturing. As stress increases beyond the ductile material's elastic limit (as defined by its yield strength), plastic deformation occurs. This plastic deformation is often accompanied by the phenomenon known as strain hardening. That is why rupture takes place at the ductile material's ultimate strength, which is greater than its yield strength. Engineers and materials scientists define *malleability* as the ability of a material to deform under compression. Gold is very malleable and ductile, so artisans and goldsmiths can hammer and roll this precious metal into extremely thin sheets. They can also pull gold into extremely thin wires.

Since brittle materials such as ceramics and concrete do not experience appreciable plastic deformation, they do not undergo the phenomenon of strain-hardening. These materials have the same value for ultimate strength and yield strength. A brittle material usually fractures when the applied strain exceeds the substance's ultimate strength. Glass also fails in this manner. When research technicians carefully pick up and reassemble the pieces of test glass intentionally shattered during experiments, they typically observe that the broken pieces fit together quite well and exhibit no signs of plastic deformation.

The behavior of solid materials under stress and strain is a very complex subject. Scientists are constantly improving their understanding of how substances behave in order to construct better, safer products. If a ductile metal is bent (flexed) too many times—even staying within its elastic limit—a deformation called *metal fatigue* can appear. After many flexure-induced microscopic distortions occur within the metal, tiny cracks begin to appear on the surface, indicating the approach of a potential catastrophic failure. Any engineered system in which metal components are subjected to cyclic episodes of deformation can experience the phenomenon of metal fatigue. Good engineering design, improved materials, routine maintenance, and regularly scheduled inspections can help prevent undesirable failures.

SOME IMPORTANT CONCEPTS ASSOCIATED WITH HEAT

Heat transfer and thermodynamics play major roles in understanding the behavior of all materials, including solid substances. This section introduces some of the basic concepts used by scientists to explain the nature of heat and the thermodynamic behavior of materials. Unfortunately, some terms, such as *heat capacity* and *latent heat,* are potentially confusing. These terms appeared in the 18th century during the development of now-discarded theories of heat, yet they remain as part of today's scientific lexicon because of historic tradition.

From the late 19th century on, scientists no longer viewed heat as an intrinsic property of matter, as was erroneously done in phlogiston theory, nor as a special, conserved fluid, as was incorrectly assumed in caloric theory. Rather, they began to recognize that heat, like work, is as a form of energy in transit. On a microscopic scale, it is sometimes useful to envision heat as disorganized energy in transit—that is, the somewhat chaotic processes taking place as molecules or atoms randomly experience more energetic collisions and vibrations under the influence of temperature gradients.

Heat transfer plays an important role in materials science. There are three basic forms of heat transfer: conduction, convection, and radiation.

Conduction is the transport of heat through an object by means of a temperature difference from a region of higher temperature to a region of lower temperature. The primary mechanism by which heat travels (conducts) through an object is the motion (vibration) of atoms and molecules. In liquids and gases, thermal conduction is due to molecular collisions. For solids, atomic vibrations are generally responsible for heat conduction. Higher temperature molecules and atoms vibrate in place more energetically, thereby transferring energy to adjacent molecules or atoms in lower temperature regions of the solid. For electrically conducting solids and liquid metals, thermal conduction is primarily accomplished by the migration of fast-moving electrons.

Convection is the form of heat transfer characterized by mass motions within a fluid, resulting in the transport and mixing of the properties of that fluid. One well-recognized example of convection is the up- and downdrafts in a fluid that is being heated from below while in a gravitational environment. Because the density of the heated fluid is lowered, the warmer fluid rises (natural convection); after cooling, the density of the fluid increases, and it tends to sink.

Finally, *radiation* (or radiant) heat transfer involves transfer of energy by the electromagnetic radiation that arises due to the temperature of a body. For objects found on Earth, most energy transfer of this type occurs in the infrared portion of the electromagnetic spectrum, but if the emitting object has a high enough temperature, it also will radiate in the visible portion of the spectrum and beyond. Scientists often use the term *thermal radiation* to distinguish this form of electromagnetic radiation from other forms, such as radio waves, light, X-rays, and gamma rays. Unlike convection and conduction, radiation heat transfer can take place in and through a vacuum.

There are two very important concepts associated with radiation heat transfer. First, every material substance in the universe has an absolute temperature (expressed in Rankines [R] or kelvins [K]) associated with its energy content. Second, every material object radiates energy away at a rate proportional to the fourth power of its absolute temperature. Toward the end of the 19th century, the Austrian physicists Josef Stefan (1835–93) and Ludwig Boltzmann (1844–1906) collaborated on the formulation of an important physical principle, now called the Stefan-Boltzmann law. This physical law states that the energy radiated away per unit of time by a blackbody is proportional to the fourth power of the object's absolute temperature. The blackbody represents the perfect emitter (and perfect absorber) of electromagnetic radiation. In nature, not every object behaves like a blackbody, but objects often approximate this behavior over certain temperature ranges and for certain radiating surface conditions. Radiation heat transfer is a very complex phenomenon, and further elaboration exceeds the scope of this section.

Thermal conductivity (symbol *k*) is a measure of a substance's ability to transport heat by means of conduction. Typical American customary units are expressed as Btu/(hr-ft-°F) and typical SI units are expressed as J/(s-m-°C). Metals such as copper and silver have high values of thermal conductivity, while foam materials have very low values of thermal conductivity. The thermal conductivity of silver at 80°F (27°C) is 248 Btu/(hr-ft-°F) (429 J/[s-m-°C]). In contrast, the thermal conductivity of urethane rigid foam (insulation) at 80°F (27°C) is just 0.015 Btu/(hr-ft-°F) (0.026 J/[s-m-°C]). Clearly, some solid materials, called *thermal conductors,* transport energy as heat very well; other materials, called *thermal insulators,* resist the conduction of heat.

In the 18th century, the Scottish physician and chemist Joseph Black (1728–99) introduced the concepts of specific heat and latent heat. Even though Black endorsed the erroneous phlogiston theory of heat, his insights into how different substances experienced different increases in temperature when they absorbed the same quantities of heat were quite accurate. Black also recognized that a solid substance required a special amount of heat addition (which he called the latent, or hidden, heat), while experiencing a change in state, when heated and completely melted at constant pressure under isothermal (constant temperature) conditions.

Scientists use *specific heat* (symbol *c*) in heat transfer analysis and thermodynamics as a measure of the heat capacity of a substance per unit mass. They define specific heat as the quantity of heat necessary to raise the temperature of a unit mass of substance by one degree (°F [°C]). At a temperature of 80°F (27°C) and a pressure of one atmosphere, silver (Ag) has a specific heat of 0.057 Btu/(lbm-°F) in American customary units or 0.234 kJ/(kg-°C) in SI units.

One practical use of the specific heat of a substance is to determine how much energy must be added as heat to a given mass of that substance to change its temperature from some reference state (called T_0) to a higher temperature state (called T_1). The process is called *sensible heat* addition when the substance being heated does not undergo a change in state. Consider a 2.2 lbm (1 kg) mass of silver (m_{Ag}) initially at room temperature and one atmosphere pressure. Using the substance's specific heat, an engineer can calculate the amount of heat required (symbolized as ΔQ_{needed}) to raise the metal's temperature from T_0 = 80°F (27°C) to T_1 = 117°F (47°C). The engineer uses the following simple equation: $\Delta Q_{needed} = c_{Ag} m_{Ag} (T_1 - T_0)$, where c_{Ag} is the specific heat of silver. After inserting appropriate values, the engineer determines that it will take about 4.55 Btu (4,800 joules) to raise the block of silver's temperature by 37°F (20°C).

Different materials have different values of specific heat. The specific heat of common red brick is about 0.20 Btu/(lbm-°F) or 0.84 kJ/(kg-°C) at a temperature of 80°F (27°C) and a pressure of one atmosphere. Using the same equation, the engineer calculates that it would take approximately 16 Btu (16,800 J) to raise the temperature of a 2.2 lbm (1 kg) mass brick by 37°F (20°C).

Not every time a solid is heated does its temperature rise. If the material is at its melting point, the addition of heat will cause the material to experience a change of state. Scientists define the *melting point* as the temperature (at a specified pressure) at which a substance experiences a change from the solid state to the liquid state. At this temperature, the solid and liquid states of a substance can exist together in equilibrium. At one atmosphere pressure, the approximate melting points (expressed in absolute temperature units) of some solid substances are: sulfur, 705.6 R (392 K); tin, 909 R (505 K); lead, 1,082 R (601 K); aluminum (pure), 1,679 R (933 K); silver, 2,223 R (1,235 K); gold, 2,405 R (1,336 K); copper (pure), 2,444 R (1,358 K); iron (pure), 3,258 R (1,810 K); titanium, 3,515 R (1,953 K); and tungsten, 6,624 R (3,680 K).

While investigating the transformation of solids into liquids (by melting) and liquids into gases (by boiling), Joseph Black coined the term *latent heat* to describe the heat energy flow in (or out) of a substance necessary to achieve the physical change of state. This term lingers in thermodynamics but has largely been replaced by the general term *heat of transformation* (symbol *L*) or the more rigorous term *enthalpy of transformation*. (Enthalpy [symbol *H* or *h*] is a somewhat complicated thermodynamic property of the substance or system under study.) The heat of transformation is the amount of heat per unit mass that must be added to (or removed from) a substance in order to produce a constant temperature (isothermal) change in physical state. Scientists define *melting* as the process whereby a solid substance transforms into a liquid. The *heat of melting* is the amount of heat that must be added to a unit mass of solid in order to change it completely into a liquid.

Scientists call the reverse process *freezing*. The *heat of fusion* is the amount of heat that must be removed from a liquid while it changes (freezes) into a solid. Melting is an exothermic process, meaning heat must be added to the solid in order to give its molecules or atoms more energy to escape their rigid structure and move about more easily. Freezing is an endothermic process, meaning energy must be removed from a substance in the liquid state in order to slow down its atoms and molecules and allow them to form a rigid structure. With a melting point of 491 R (273 K), more commonly expressed as 32°F (0°C), ice has a heat of melting of 143 Btu/lbm (333 kJ/kg) at one atmosphere pressure. The heat of melting of lead is 9.98 Btu/lbm (23.2 kJ/kg); silver, 45.2 Btu/lbm (105 kJ/kg); and copper, 89 Btu/lbm (207 kJ/kg).

During isothermal (constant temperature) changes in state, the heat of fusion and the heat of melting have the same numerical values. What is different is the direction of heat flow into (for melting) or out of (for freezing) the substance. The term *melting point* is synonymous with the term *freezing point.* A liquid freezes when the necessary amount of heat is removed at the freezing point and the substance transforms into the solid state.

Certain substances undergo a transition directly from the solid state into the gaseous (vapor) state without passing through the liquid state. Scientists call this process *sublimation.*

At one atmosphere pressure, solid carbon dioxide (commonly called *dry ice*) sublimes directly into gaseous carbon dioxide at a temperature of –109.6°F (350.4 R) or –78.5°C (194.65 K). The corresponding heat of sublimation is 245.5 Btu/lbm (571 kJ/kg). Scientists define *deposition* as the reverse process, in which a gas directly transforms into a solid without passing through the liquid state. Sublimation requires that heat be added to the solidified gas to transform it directly into a vapor; deposition requires that the gas have heat removed in order to transform it directly into a solid. Under the proper atmospheric conditions, water vapor can directly transition into the solid state and appear as frost or snowflakes.

ELECTROMAGNETIC PROPERTIES OF SOLID MATTER

This section briefly discusses one of the more interesting and important properties of matter, namely electromagnetism. From telecommunications to computers, from electronic money transfers to home entertainment centers, the application of electromagnetic phenomena dominates civilization this century.

The history of magnetism extends back into antiquity. The English word *magnet* traces its origin to the ancient Greek expression "stone of magnesia" (μαγνητης λιθος). The reference is to an interesting type of iron ore named *lodestone* by alchemists that is found in Magnesia, a region in east central Greece.

Modern scientists now say that each of an atom's electrons has an orbital magnetic dipole moment and a spin magnetic dipole moment. A material exhibits magnetic properties when the combination of each of these individual dipole moments produces a net magnetic field. Although a detailed, quantum-mechanical description of the phenomena that give rise to magnetism is beyond the scope of this book, the brief discussion that follows should prove helpful in understanding the basic magnetic properties of materials.

Scientists identify three general types of magnetism: ferromagnetism, paramagnetism, and diamagnetism. Ferromagnetic materials, such as iron, nickel, and cobalt, have the resultant magnetic dipole moments aligned and consequently exhibit a strong magnetic field. These substances are normally referred to as magnetic materials or ferromagnetic materials. Quantum physicists suggest that a ferromagnetic substance enjoys the cooperative alignment of the electron spins of many atoms.

Materials containing rare earth elements, transition elements, or actinoid elements exhibit a property called paramagnetism. The magnetic dipole moments within such materials are randomly oriented. As a result, there is no net alignment that combines to create a significant magnetic field. However, in the presence of a strong external magnetic field, the individual magnetic dipole moments in a paramagnetic material tend to align and become weakly attracted by the external magnetic field. Quantum physicists suggest that this attraction is generally the result of unpaired electrons. The magnetic field of a paramagnetic substance, while observable, remains weaker than the magnetic field exhibited by ferromagnetic materials. Lithium, magnesium, molybdenum, and tantalum are examples of paramagnetic materials.

Finally, most of the elements in the periodic table (including copper, gold, and silver) exhibit the property of diamagnetism. The atoms of diamagnetic substances produce only weak magnetic dipole moments, so even when the substance is placed in a strong external magnetic field, the diamagnetic material exhibits only a feeble net magnetic field (if any). The diamagnetic substance is not attracted by an external magnetic field, or else the substance may be slightly repelled by this field. Quantum physicists suggest that diamagnetic substances have only paired electrons.

Today, physicists measure the strength of a magnetic field *(**B**)* with a unit called the *tesla* (T). One tesla is equal to one newton per ampere per meter, that is, $1 \text{ T} = 1 \text{ N A}^{-1} \text{ m}^{-1}$. The tesla is a derived SI unit named in honor of the Croatian-born Serbo-American electrical engineer and inventor Nikola Tesla (1870–1943). Scientists also use an earlier, non-SI unit for a magnetic field called the *gauss* (G). This unit honors the German mathematician and scientist Carl Friedrich Gauss (1777–1855). The two units of magnetic field are related as follows: $1 \text{ T} = 10^4 \text{ G}$. The magnetic field near Earth's surface is about 10^{-4} T, or 1 G. A powerful electromagnet produces a magnetic field of 1.5 T, or 1.5×10^4 G. Finally, physicists express magnetic flux *(**Φ**)* in terms of webers (Wb), where $1 \text{ Wb} = 1 \text{ T m}^{-2}$. The weber is a derived SI unit that honors the German physicist Wilhelm Eduard Weber (1804–91).

A major breakthrough in the physics of electromagnetism occurred quite serendipitously. On April 21, 1820, the Danish physicist Hans Christian Ørsted (1777–1851) was making preparations for separate laboratory demonstrations about electricity and magnetism. A professor of science at the University of Copenhagen, Ørsted was using the newly invented Voltaic pile (an early electric battery) as his source of electricity. As he moved the equipment around, he noticed that a current-carrying wire caused an unexpected deflection in the needle of a nearby compass. By coincidence, that particular wire was carrying a current because he was planning to discuss how the flow of electricity from a Voltaic pile could cause heating in a wire. The fortuitous deflection of the compass would electrify the world.

Although Ørsted could not explain how the flow of electricity from a battery could cause this magnetic effect, he felt obliged, as a scientist, to report his unusual observations to Europe's scientific community. His discovery clearly linked electricity and magnetism for the very first time. Once announced, Ørsted's findings immediately encouraged other scientists, such as the French physicist André-Marie Ampère (1775–1836), to launch their own comprehensive investigations of electromagnetism. The overall technosocial impact of this new emphasis on electromagnetism by scientists in the early 19th century was the emergence of the electricity-based Second Industrial Revolution.

The German physicist George Simon Ohm (1787–1854) published the results of his experiments with electricity in 1827. His research suggested

THE FIRST CHEMICAL BATTERY

His professional disagreement with Luigi Galvani (1737–98) concerning the nature of animal electricity encouraged Count Alessandro Volta (1745–1827) to perform additional experiments involving the electric nature of matter. In 1800, Volta developed the voltaic pile—the first chemical battery. From a variety of experiments, Volta determined that in order to produce a steady flow of electricity he needed to use silver and zinc as the most efficient pair of dissimilar metals. First, he made individual cells by placing a strip of zinc and silver into a cup of brine. He then connected several cells to increase the voltage. Finally, he created the first voltaic pile (chemical battery) by alternately stacking up discs of silver, zinc, and brine-soaked heavy paper—quite literally in a pile. Soon, scientists all over Europe used and improved Volta's invention to give themselves a steady, dependable flow of electricity (direct current) for their experiments. Volta's battery enabled the scientific study of electromagnetism in the 19th century.

the existence of a fundamental relationship between voltage, current, and resistance. In particular, Ohm stated that the electrical resistance (R) in a material may be defined as the ratio of the voltage (V) applied across the material to the electric current (I) flowing through the material, or $R = V/I$. Today, physicists and engineers call this important relationship *Ohm's law.*

In recognition of his contribution to the scientific understanding of electricity, scientists call the SI unit of resistance the ohm (symbol Ω). One ohm of electric resistance is defined as one volt per ampere ($1\ \Omega = 1\ V/A$). Scientists originally called the reciprocal of the ohm the *mho* and used it as the unit of electrical conductance. Today, scientists call the SI-derived unit of electrical conductance the siemens (symbol S) in honor of the German engineer and industrialist Ernst Werner von Siemens (1816–92).

The British experimental physicist Michael Faraday (1791–1867) and his American counterpart Joseph Henry (1797–1878) independently discovered the physical principles behind two of the most important electric-powered machines in modern civilization: the electric generator and the electric motor. Later in the 19th century, the Scottish scientist James Clerk Maxwell (1831–79) presented a comprehensive set of equations that theoretically described electromagnetism. Maxwell's work revolutionized both physics and the practice of engineering. Inventors such as Thomas Edison (1847–1931) and Nikola Tesla applied electromagnetism to numerous new devices, providing people surprising new power and comforts. The discoveries about the nature of electromagnetism during the 19th century reached an exciting scientific climax in 1897. In that year, the British physicist Sir J. J. (Joseph John) Thomson (1856–1940) published the results of his experiments that clearly demonstrated the existence of the electron, the first known subatomic particle.

Today, scientists recognize that some substances, such as silver and copper, readily conduct the flow of electricity and call these materials *electrical conductors.* Other materials, such as wood and most plastics, resist the flow of electrons and are called *electrical insulators.* The basic difference between conductors and insulators lies in their atomic structures. A material that is a good electrical conductor has one or several outer (valence) electrons that are loosely attached to the parent atomic nucleus and are therefore available to freely wander through the material under the influence of an applied voltage. The situation is quite different for a material that is a good electrical insulator. In this case, almost every electron remains tightly bound to its parent atom, so that even when a voltage difference is applied across the material, the outer electrons cannot wander freely through it.

When scientists discuss the property of electrical conductivity, they often include a third common type of material called a *semiconductor.* A semiconductor is a solid crystalline material, such as silicon (Si) and germanium (Ge), that has a typical electrical conductivity intermediate between the values of good electrical conductors and insulators. Semiconductors can carry electric charges, but not very well. The conductance of silicon is about 1 million times less than the electrical conductance of copper.

With an electrical conductivity (symbol σ) of 63×10^6 siemens per meter (S/m) at a temperature of 540 R (300 K), silver is an excellent conductor of electricity. Copper follows closely behind, with an electrical conductivity of about 60×10^6 S/m at 540 R (300 K). For economic and resource availability reasons, copper is widely used in the distribution of electricity. Gold has an electrical conductivity of 45×10^6 S/m at 540 R (300 K), and aluminum, about 38×10^6 S/m at 540 R (300 K). Note that the electrical conductivity of metals is temperature dependent and decreases significantly with increasing temperature.

The electrical resistivity (symbol ρ) of a material is the reciprocal of that material's electrical conductivity. Silver has an electrical resistivity of 1.6×10^{-8} ohm-meter (Ω-m) at a temperature of 540°R (300 K). At 540°R (300 K), pure silicon has an electrical resistivity value of 2,500 Ω-m. A typical n-type silicon semiconductor material has an electrical resistivity of 8.7×10^{-4} Ω-m, and a typical p-type silicon semiconductor material has an electrical resistivity of 2.8×10^{-3} Ω-m. Engineers call silicon that has been doped with (electron) donor atoms, such as phosphorus, *n-type semiconductors;* silicon that has been doped with (electron) acceptor atoms, such as aluminum, are called *p-type semiconductors.* Finally, they regard glass and fused quartz as good electrical insulators. The electrical resistivity of glass at 540 R (300 K) ranges from 10^{10} to 10^{14} Ω-m. Fused quartz has an approximate electrical resistivity of 1×10^{16} Ω-m.

The digital (or information) revolution is a generic expression that actually encompasses several major technology shifts that occurred in the mid- to late 20th century. The first part of this process was the discovery of the transistor in the late 1940s; the second part involved the application of the transistor in the subsequent microelectronics revolution. The integrated circuit was nothing short of a technical miracle in materials science and engineering. This device accelerated the exponential development and application of digital computers and microprocessors.

OPTICAL PROPERTIES OF SOLID MATTER

How electromagnetic (EM) radiation, including visible light, interacts with solid matter involves some of the more fascinating and important areas of materials science. As discussed here, radiation can be scattered by the surface of a solid material, absorbed by the material, or transmitted by the material.

The study of the optical properties of solid matter is an enormously large and complex field. This section introduces only the most basic and essential concepts in *optics,* the branch of physics that involves the manipulation and analysis of EM radiation. Scientists define an *optical device* as a device that allows them to manipulate, alter, or control EM radiation. A lens is a transparent substance designed to transmit, bend, and focus radiation of a certain wavelength region. A shiny, silver-surfaced mirror is a simple optical device that reflects visible light. A solid-state laser is a sophisticated device that concentrates and amplifies radiation. The term *laser* is an acronym for *l*ight *a*mplification by *s*timulated *e*mission of radiation. At the heart of a solid-state laser is a crystal, such as a ruby, whose atoms interact with light in such a way as to create a very intense, well-collimated beam. Since the 1960s, lasers of all types have influenced modern life. The optical scanner at the supermarket, the laser light pointer used during a college professor's lecture, and the medical laser used by a surgeon are some examples.

In optics, scientists apply the conservation-of-energy principle to a narrow beam of light that impinges upon a slab of material of a specified thickness. The incident beam of light experiences reflection, absorption, and/or transmission. Sometimes the solid material reflects most or all of the incoming light. In this case, the solid material is either a diffuse reflector or a specular reflector. Scientists call any surface that reflects incident light in a multiplicity of directions a diffuse reflector. Rough surfaces are typically diffuse reflectors. The opposite is a mirrorlike solid surface that functions as a specular reflector. In specular reflection, the angle of the incident light beam is equal to the angle of the reflected light beam as measured with respect to the normal (perpendicular) to the surface.

In addition to reflection, an incident beam of light can enter a solid material and pass entirely through it (transmission) or be completely absorbed within the solid material (absorption). Considered together, the three processes (reflection, absorption, and transmission) account for all the photons (energy) in the incoming beam of electromagnetic radiation. Most materials are not perfect reflector, absorber, or transmission media. EM radiation–solid matter interaction processes generally involve a combination of all three physical phenomena in varying degrees. A *transparent material* allows most of the beam of radiation to pass through without experiencing much reflection or absorption. A *translucent material* allows much of the incident beam to scatter within, while the remaining light is transmitted or reflected. An *opaque material* is one that absorbs the entire incident beam, such that none of the light photons pass through the substance. A pane of window glass is transparent to sunlight, a stained glass window is translucent, and the glasslike mineral obsidian is opaque.

How a solid absorbs radiation is a complicated process that depends upon microscopic phenomena such as chemical bonding and atomic structure. The wavelength of the incident EM radiation is another important physical parameter. Some solids are excellent absorbers of infrared radiation but relatively poor absorbers of visible radiation.

Spectroscopy is the study of the spectral lines from various atoms and molecules in an object. Scientists use spectroscopy to infer the material composition of objects, to gather information about chemical bonding, and to determine the electronic properties of various materials. Forensic scientists use spectroscopic analysis to identify minute traces of unknown materials collected at a crime scene. Astronomers use emission spectroscopy to infer the material composition of a celestial object that has emitted the observed light; they use absorption spectroscopy to infer the composition of the intervening medium, such as interstellar dust or a star's outer atmosphere.

Certain substances exhibit the intriguing phenomenon of fluorescence. When a person looks at fluorescent minerals in ordinary daylight, the minerals appear rather ordinary and dull. However, if the person views the same minerals under ultraviolet light (which is invisible to the human eye), characteristic, often vivid, colors suddenly appear. Scientists explain the process as follows. Higher energy (shorter wavelength) ultraviolet photons interact with the atomic electrons on the surface of the substance, exciting these electrons to higher energy states. The excited electrons then return to their normal (ground) states by emitting two or more lower-energy (longer-wavelength) photons. Since these lower-energy de-excitation photons occur in the visible portion of the spectrum, they create the vivid color display. Scientists use fluorescence as an important tool in identifying substances.

CHAPTER 5

Rocks and Minerals

This chapter describes how the rocks and minerals found on Earth experience a cycle of continuous breakdown and rebirth. Scientists refer to this process, which is driven in part by plate tectonics, as the rock cycle. Selected minerals and natural gemstones are also discussed.

THE SOLID EARTH

In Latin, the phrase *terra firma* means solid earth. Geophysicists speak of solid Earth science, and astronomers place the eight major planets of the solar system into two general categories: large, low-density, gaseous worlds, such as Jupiter, that do not have solid surfaces and small, high-density, rocky worlds, such as Earth, that have solid surfaces. As will become apparent in this section, the notion of a solid Earth, while psychologically comforting to many, is actually a significant oversimplification.

Although Earth has a huge variety of surface features, scientists can generally explain their origins by using just four major geological processes: impact cratering, volcanism, tectonics, and erosion. Impact cratering is the process in which an asteroid or comet hits (impacts) a planet's solid surface, carving out a large bowl-shaped crater as a result of this very energetic cosmic collision. Volcanism involves the eruption of molten rock onto the planet's surface from its interior. Scientists call this molten rock *magma* when it resides in the planet's interior and *lava* when it emerges and flows on the planet's surface. Tectonics describes the disruption of a planet's surface by internal stresses. Heat-driven convection within Earth's mantle has fractured the lithosphere into more than a dozen pieces, called plates. Finally, erosion involves the construction (building up) or destruction (wearing down) of surface features by the actions of wind, water, ice (glacial movements), and other environmental phenomena.

Earth is a restless planet that evolved during the past 4.6 billion years, transitioning from a totally molten world into a habitable planet that possesses a solid outer skin. The surface of humans' home planet continues to be influenced by the Sun, gravitational forces, complex interactions with the hydrosphere and atmosphere, and processes emanating from deep within its core. On very short time scales, people appear to be standing on solid ground, or terra firma. Despite such stable appearances, Earth continuously experiences many processes that transform and sculpt its surface. These Earth-changing processes can be quite dramatic and even lead to loss of life as well as extensive property damage. Sometimes these changes happen slowly, such as surface subsidence due to aquifer depletion. Other times the changes are unpredictable and rapid, such as earthquakes, volcanic eruptions, and landslides.

Earth's surface and interior are major components of an interconnected, highly dynamic physical system. Variations in the ice sheets and land cover impact climate and the environment. Violent events such as earthquakes, volcanic eruptions, landslides, and floods reshape the surface and pose significant hazards. Solid-earth scientists measure both the slow and fast deformations of the planet's surface in order to improve their overall understanding of the dominant geological processes. This knowledge helps them optimize responses to natural hazards and identify potential risk areas.

Earth has an equatorial radius of 3,963 miles (6,378 km). Starting at the surface and descending downward (the direction of increasing pressure, temperature, and density), the planet consists of three distinctive layers: crust, mantle, and core. Scientists find it convenient to divide these layers according to their respective densities. The crust forms the rigid but thin, rocky outer skin of the planet. It is about 31 miles (50 km) thick or less and contains low-density rock materials, such as granite and basalt. Below the crust lies a thick layer (about 1,800 miles [2,900 km] in depth) called the mantle. The mantle consists of hot, rocky material of modest density and contains minerals mixed with silicon, oxygen, and other elements. Below the mantle is the high-density spherical core, consisting of an outer (liquid) region (about 1,400 miles [2,260 km] thick) and an inner (solid) region (about 727 miles [1,170 km] in radius)—each of which contains metals, especially iron and nickel.

The outermost layer of the mantle is solid and along with the crust forms a region scientists define as the lithosphere. *Lithos* (λιθος) is the Greek word for stone. The lithosphere is about 93 to 124 miles (150 to 200 km) thick under the continents but less under the oceans. The crust and upper mantle are subjected to the dynamic forces produced by mantle convection, which causes segments of the shell and materials at the top to break up into plates. The term *tectonics* refers to any planetary surface changes that result from the compression (squeezing), stretching (tension), or other forces acting upon the lithosphere. Scientists define *plate tectonics* as the geological process whereby plates move under, over, and around each other in the lithosphere. When two plates collide, surface compression occurs, and mountain building takes place.

THE ROCK CYCLE

The rock cycle is the scientific model that describes the origin of the three fundamental types of rocks found on Earth: igneous, sedimentary, and metamorphic. As shown in the accompanying illustration, the rock cycle is an ongoing process that begins as rocks are pushed up to the surface by tectonic forces and then eroded by environmental forces such as wind and water. The word *igneous* comes from the Greek word for fire. Igneous rocks form when molten material (magma) in the lithosphere cools and solidifies. When this crystallization process occurs beneath Earth's surface, intrusive igneous rocks form. When there is a volcanic eruption that leads to the flow of lava onto Earth's surface, extrusive igneous rocks form. The magma itself is derived from the melting of preexisting crustal rocks or from material sources below Earth's crust. Geologists call the subsurface formation and movement of magma *plutonism* after Pluto, the god of the underworld in Greco-Roman mythology.

Scientists use the term *volcanism* to refer to the process in which magma reaches the surface and erupts. A volcano forms when molten rock and solidified volcanic debris are ejected onto the surface. In addition to rock material, volcanoes also vent large quantities of water vapor and various gases into the atmosphere. Scientists call the molten rock material that flows from a volcano onto Earth's surface *lava*. Once cooled, lava assumes various forms of extrusive igneous rock. The native Hawaiian language has many words to describe these various forms of lava. For example, aa (or block) lava solidifies as a mass of blocks and fragments characterized by a rough surface. Pāhoehoe (or corded) lava has a smooth, billowy, ropy surface.

Rocks at or near Earth's surface experience weathering and erosion. The continual and relentless action of environmental forces in the

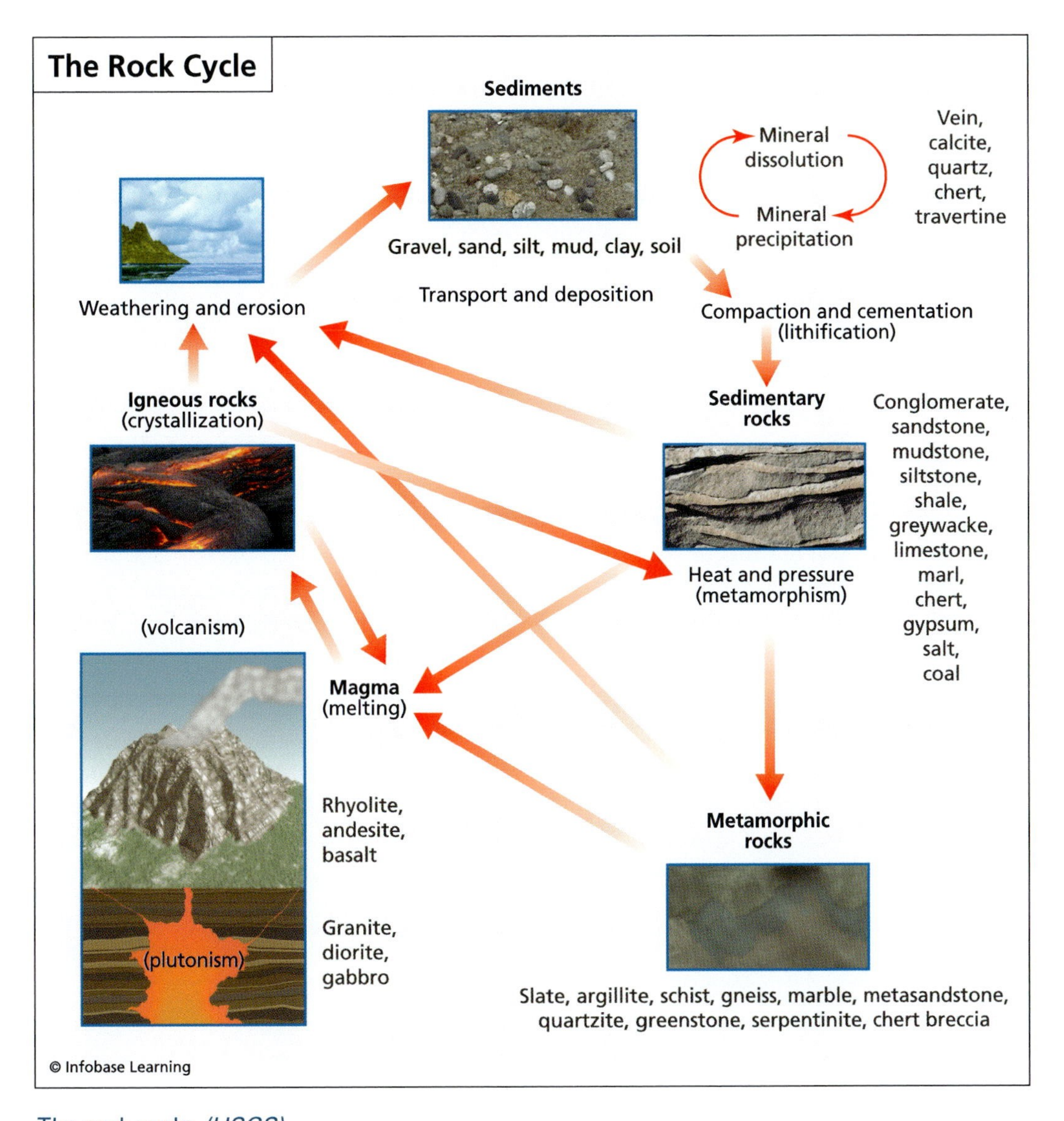

The rock cycle *(USGS)*

atmosphere (such as wind) and the hydrosphere (such as flowing water) slowly disintegrate and decompose these rocks. Rainwater often seeps into cracks in rock and then freezes. As ice forms and expands in a crack, it can exert tremendous forces on the rock—often with sufficient strength to break it. Geologists refer to these natural erosion processes as *mechanical weathering.* There are many environmental actions that contribute to converting rocks into smaller sediments as part of the overall rock cycle. *Chemical weathering* involves the breakdown of surface rocks by chemical actions. The minerals found in igneous rocks are often unstable under normal atmospheric conditions and consequently vulnerable to attack by water, especially flowing water that has become acidic or alkaline.

Geologists often describe the breakdown of surface rocks into smaller and smaller fragments using the following descriptive order, from larger to smaller fragments: boulders, cobbles, gravel, sand, silt, and microscopic particles. The finer the particle, the easier it is for natural forces to move it. Sediments are transported by water or wind erosion, and rock fragments experience the dissolution and transport of soluble chemical components by surface water and groundwater.

As part of the rock cycle, rock fragments are transported and deposited in a variety of sedimentary environments. On land, such sedimentary environments include swamps, stream flood plains, dunes, and desert basins. Along coastal regions, sediments can accumulate in the deltas at the mouths of rivers, in lagoons, on beaches, and on barrier islands. The great majority of sediments ultimately come to rest in the world's oceans. The sediments can accumulate in massive deposits that form continental shelves or else continue into the deep ocean basins beyond.

As sediments accumulate over geologic time scales, the weight of the overlying materials causes compaction of the materials in the lower sedimentary layers. In addition to compaction, sedimentary materials can experience cementation, a process in which previously dissolved and transported minerals precipitate in pore spaces and join (cement) the grains of sediment together. Geologists refer to the combined effects of compaction and cementation as the process of *lithification.* This process results in the formation of sedimentary rocks such as sandstone, chalk, and limestone.

Over time, sedimentary rocks formed near Earth's surface typically experience weathering and erosion as part of the overall rock cycle, but sedimentary rocks formed in a basin along the margin of a continent may experience increased burial (subduction) due to plate tectonics. As

HOW NATURE'S FORCES CREATED YOSEMITE NATIONAL PARK

Most people who visit Yosemite National Park in California gaze in awe upon the 2,425-foot (739 m)-high Yosemite Falls, the tallest waterfall in North America. The landscapes and features of this beautiful park inspire wonder at the natural forces that combined to create such majesty. Scientists have carefully investigated and documented Yosemite's landscapes in order to provide an explanation. Their efforts suggest that Yosemite National Park is a glaciated landscape and that its beautiful scenery results from the interaction of ancient glaciers and the underlying rocks.

Yosemite National Park's geology is intimately linked to the origin of the Sierra Nevada, a mountain range that stretches along California's eastern flank. It may be hard to believe, but what are now the Sierra Nevada and the Central Valley of California were once under water. As depicted in the accompanying

(continues)

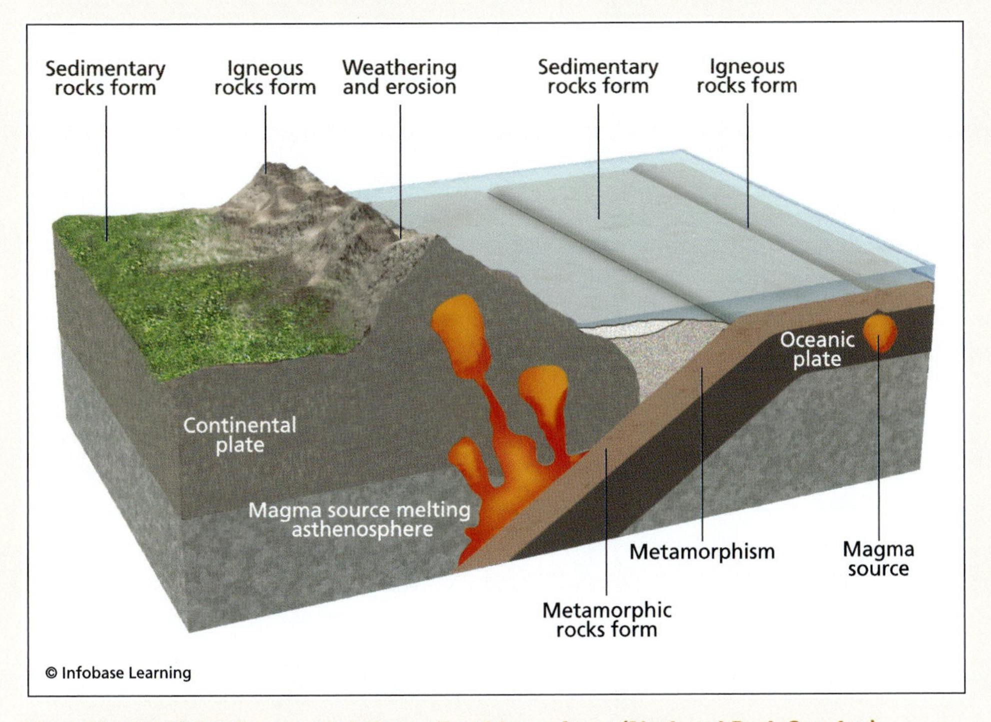

Great forces that change and shape Earth's surface ***(National Park Service)***

(continued)

illustration, large quantities of sand, silt, and mud eroded from primordial mountain ranges surrounding an ancient sea and settled to the seabed in layers. Sedimentary rocks slowly formed as part of this geologic process. Then, great forces deep within Earth's crust warped these layers of sedimentary rock and folded them into a mountain range that extended from northeast to southwest.

These geologic forces also changed sedimentary rock into metamorphic rock. As the mountains rose, pockets of magma from the asthenosphere began to form beneath them. Geologists define the *asthenosphere* as the region in Earth's upper mantle just below the lithosphere. This region contains semi-molten rock material that is chemically similar to the rock material found in the overlying lithosphere. The molten rock eventually cooled and solidified into the granitic rocks found today in the national park. Over millions of years, weathering and erosion stripped away the overlying metamorphic rocks and sculpted a landscape of meandering streams, rolling hills, and broad valleys. Yosemite's present landforms result from geologically recent glaciations. These surface features include U-shaped canyons, jagged peaks, rounded domes, and waterfalls. Glacially polished granite is also common throughout the park. The cliffs of Yosemite Valley and many of the park's higher peaks are a testament to the durability of granite.

descending sedimentary rocks sink deeper into Earth's crust, higher temperatures and the increased lithostatic pressure can alter the physical structure and chemical composition of the original rocks, called protoliths. Geologists refer to this process as *metamorphism* and the resultant new rock as metamorphic rock. Metamorphic means "change of form." The overall process is basically a solid state change in mineralogy. For example, during metamorphism, limestone, a sedimentary rock, becomes marble, a metamorphic rock, as a result of the increased heat and pressure. Igneous rocks, sedimentary rocks, and even preexisting metamorphic rocks can experience metamorphism.

Scientists regard the presence of specific minerals in a metamorphic rock as indicative of the degree of heat and pressure the original protolith has experienced. Generally, protoliths buried in Earth's crust are changed by the heat and pressure they encounter from their original physical and chemical conditions into metamorphic rocks with new minerals and structures. Metamorphism is a complicated process, and a more detailed discussion is beyond the scope of this section.

CHARACTERISTICS OF MINERALS

Scientists define a *rock* as an aggregate of one or more minerals. The term *aggregate* suggests that the minerals are bound together in a mixture in which each mineral retains its properties. As mentioned earlier, there are three basic types of rock: igneous, sedimentary, and metamorphic. Scientists usually define a *mineral* as a naturally occurring inorganic solid that possesses a specific chemical composition and a definite internal structure. In practice, this definition is not always applied with scientific precision. Geologists typically classify coal as a mineral, despite the fact that the hydrocarbon is actually an organic compound.

Like all matter, minerals consist of elemental atoms. Some minerals, such as gold, consist entirely of atoms of just one element. Other minerals, such as halite (rock salt), represent a combination of two or more chemical elements—here, sodium (Na) and chlorine (Cl) to form the crystalline mineral sodium chloride (NaCl), more commonly known as table salt. Geologists and mineralogists often use a dimensionless number called the *specific gravity* to characterize various minerals and gemstones. They define specific gravity as the substance's density divided by the density of water. The reference density of water is usually taken as 62.4 lbm/ft^3 (1 g/cm^3), a density value corresponding to the properties of water at one atmosphere pressure and 39°F (4°C) temperature. If an object has a specific gravity greater than 1, it is denser than water.

It is important to recognize that minerals are solid, crystalline structures of inorganic substances characterized by an orderly array of atoms held together by chemical bonds. Some chemical elements are able to join together in more than one way. The term *polymorph* means "many forms." Scientists use this term to identify two minerals that have the same chemical composition but totally different physical properties. The pure carbon minerals diamond and graphite are excellent examples of polymorphs. Although both minerals consist of elemental carbon, diamond is the hardest known natural substance, while graphite is a soft gray-black substance used in pencils and as a dry lubricant. (See chapter 8.)

When struck with a tool such as a wedge, some minerals, such as diamond and mica, will break, or cleave, along specific crystal planes as determined by their mineral structure and chemical bonding. Minerals such as quartz that do not exhibit the property of cleavage often fracture in an irregular manner when struck.

Geologists define the *hardness* of a mineral as the solid substance's resistance to scratching or abrasion. Friedrich Mohs (1773–1839) was a

SALT—PILLAR OF CIVILIZATION

Salt is a universal commodity that is consumed by virtually every human being and animal on Earth. While people are usually not familiar with the names of most of the ferrous, nonferrous, and industrial minerals that sustain their global civilizations, they easily recognize salt because of its distinctive taste and daily usage. In mineralogy, salt is called *halite,* a term that derives from hals ('αλς), the Greek work for salt. Salt is found almost everywhere in the world either in underground salt domes or in evaporated beds surrounding ancient or existing landlocked seas.

Although now a very common mineral, at one time salt was regarded as precious and as important as gold by many early societies. The history of salt production and consumption coincides quite well with the history of civilization. Prehistoric humans obtained the salt their bodies needed primarily from the meat of the animals they hunted. Many of the prehistoric animals hunted for food were more easily caught by Stone Age hunters because the animals tried to satisfy their constant cravings for salt by congregating around salt licks or saline springs.

When agriculture emerged in the Neolithic Revolution, people had to supplement the vegetables and cereals in their diets with extra quantities of salt. At this point, the direct quest for salt became an integral part of human history. Neolithic peoples soon discovered that salt could preserve foods such as meat, fish, cheese, and vegetables. Until more modern food preservation techniques (refrigeration, canning, and freezing) became available, the ability to preserve food was salt's most important application. Ham and bacon are examples of salt-preserved meats that are still in common use. The presence of salt prevents rotting by bacteria, so people soak meats and fish in salt water (brine) for a few days prior to sun-drying to assist in preservation and flavoring.

During the Bronze Age, societies began searching for supplies of salt. The two main historic sources of salt were seawater and rock salt. Salt beds that could be effectively mined were especially valued. As the Roman Empire emerged, stable supplies of salt were critical to its stability and growth. The

German geologist and mineralogist. In 1812, he devised the widely used comparative scale called the Mohs scale of mineral hardness. He based his relative hardness scale on the ability of one mineral to scratch or mark another mineral. According to this scale, talc is the softest mineral and

Romans constructed special roads to ensure a steady supply of salt to their capital city. The Via Salaria went from Rome to salt-producing evaporative coastal ponds on the Adriatic Sea. In Latin, *sal* means "salt." The English word *salary* comes from the Latin word *salarium,* meaning "payment in salt." According to historic tradition, at certain times Roman soldiers were paid a portion of their wages in salt. Since most poor people needed this mineral to supplement their diets, Roman officials controlled the price of salt to raise money and to exercise political control over the population at large.

There are many other historic examples of salt playing a major role in the location of cities, the development of international commerce, and the rise and fall of empires. Salt mines, especially the Wieliczka Salt Mine, played a major role in the political and economic development of Poland. Starting with Casimir I in about 1044, Polish monarchs recognized the great value of this important mineral resource. In the 14th century, profits from salt were generating more than 30 percent of the Polish state's income. By the 16th century, the Wieliczka Salt Mine had become one of the largest economic enterprises in all of Europe. Although the price of salt and production techniques have changed dramatically since the Middle Ages, the Wieliczka Salt Mine still generates revenue—now as a major tourist attraction, with more than 1 million visitors annually. In 1978, the United Nations Educational, Scientific and Cultural Organization (UNESCO) included the famous salt mine in its international list of cultural and natural heritage sites—sites from around the world considered to be of outstanding value to humanity.

Historians who specialize in the American colonial period often highlight the strong link between cod fishing, salt, and the economic viability of the New England colonies. When the British Parliament attempted to curb the American fishing enterprise, the political counterreaction of the New England colonists fanned the flames of revolution. Furthermore, fishermen and fish merchants from Massachusetts formed the nucleus of the fledgling U.S. Navy. Salt cod remains very popular to the present day.

has a Mohs hardness value of 1. The scale continues in the direction of increasing hardness as follows: gypsum (Mohs hardness 2), calcite (3), fluorite (4), apatite (5), feldspar (6), quartz (7), topaz (8), corundum (9), and diamond (10), the hardest known natural substance.

Other physical properties of minerals that scientist find useful include transparency, luster, and magnetism. The property of transparency refers to the ease by which visible light passes through a sample of material. A transparent mineral such as quartz allows light to pass, and objects can be clearly seen when observed through the substance. A pane of clear glass in a typical window is a common example of a transparent substance. A translucent mineral allows some light to pass through it, but objects on the other side of the material are not discernable. The gemstone jade is an example of a translucent mineral. Finally, an opaque mineral completely blocks the passage of visible light. Hematite, the principal ore of iron, is an example of an opaque mineral.

The luster of a mineral is the physical property that describes the appearance of the mineral's surface and what happens to light when it falls on the mineral's surface. Geologists and gemologists say a diamond has an adamantine luster, which means it reflects and transmits light in a brilliant fashion. Metallic minerals such as pyrite and galena have a shiny luster similar to the appearance of polished metal. Earthen minerals such as clay (kaolinite) have a dull luster. Other terms associated with a mineral's luster include waxy (as exhibited by turquoise), pearly (talc), resinous (amber), vitreous or glassy (quartz), and silky (gypsum).

Some minerals exhibit fluorescence, and others demonstrate an internal reflection of light known as sheen. Minerals, typically those with high iron content, can also exhibit the property of magnetism. Lodestone is an intensely magnetic form of magnetite, a mineral consisting of iron oxide (Fe_3O_4).

MINERALS IN EARTH'S LITHOSPHERE

Although scientists have discovered and characterized thousands of minerals, less than 100 or so are regarded as common in Earth's crust. Silicon dioxide (SiO_2) is the most abundant crustal mineral. On an elemental basis, oxygen accounts for about 46.6 percent by mass of the composition of Earth's crust; silicon, 27.7 percent; aluminum, 8.1 percent; iron, 4.7 percent; calcium, 3.4 percent; sodium, 2.6 percent; potassium, 2.4 percent; and magnesium, 1.9 percent. The remaining percentage (about 2.6) consists of the combined mass of all other elements in Earth's crust. Scientists estimated these relative abundance data by examining the average elemental composition of various samples of igneous rocks.

Another way scientists sometimes describe the elemental composition of Earth's crust is by how many atoms of each element are found in a typical chunk of crustal material. They use a large population of atoms (say, 10,000) as the basis for comparison. This approach can be a little confusing if a person forgets that different atoms have significantly different atomic masses. Within this atom-by-atom approach, there are approximately 5,330 atoms of oxygen per 10,000-atom sample of crustal material, 1,590 atoms of silicon, 1,510 atoms of hydrogen, 480 atoms of aluminum, 180 atoms of sodium, 150 atoms of iron, 150 atoms of calcium, 140 atoms of magnesium, and 100 atoms of potassium. The remaining 370 atoms per average 10,000 atoms of crustal material represent everything else in the crust, including gold, silver, carbon, and phosphorous atoms.

Geologists and mineralogists also find it useful to characterize the minerals in Earth's crust by their chemical compositions. This approach results in such commonly encountered mineral designations as silicate minerals, oxide minerals, sulfide minerals, carbonate minerals, and halide minerals.

Silicate minerals represent the largest group of crustal minerals. They are compounds of metals combined with oxygen and silicon. Quartz is a mineral member of this group. It consists of pure silicon dioxide (SiO_2), is very abundant in Earth's crust, and has commercial value in the production of glass and use as a gemstone. Although pure quartz is colorless, chemical impurities create beautiful, naturally colored quartz crystals, many of which are cut into attractive gems. When human beings in early societies developed the rudimentary materials science expertise (primarily heating and mixing) needed to modify silicate minerals (such as sand, clay, and limestone), the results included products such as ceramic pottery, bricks, glass, and cement—materials that transformed civilization.

Feldspar is the mineral name scientists have given to a group of silicate minerals characterized by the presence of aluminum and the silica ion (SiO_4). Feldspar is the single most abundant mineral group, constituting about 50 percent of the Earth's crust. Sometimes called aluminosilicates, feldspars are silicates of aluminum that contain varying amounts of sodium, potassium, iron, calcium, or barium—or combinations of these elements. Geologists commonly find veins of feldspars that have crystallized from magmas in both intrusive and extrusive igneous rocks.

They also observe feldspars in both metamorphic and sedimentary rocks. Major commercial uses of feldspar include glassmaking and the production of ceramics.

Oxide minerals are a combination of metals with just oxygen. This group of minerals is important in mining (an extractive industry) because they represent the ores from which useful quantities of certain metals can be commercially extracted. Geologists define an *ore* as a type of rock from which metals (or gemstones) can be extracted through well-established mining techniques and refining processes. Based on metal content, geologic abundance, and global demand, some ores have much greater commercial value than others. Oxide minerals of commercial value include the iron ores hematite (Fe_2O_3) and magnetite (Fe_3O_4). Another example is corundum (Al_2O_3), a mineral with an important application as an abrasive material as well as a source of gemstones. Mineralogists and mining engineers use the term *gangue* to identify the rocks and minerals in an ore deposit that have no economic value. The removal and disposal of gangue increases the cost of mineral extraction and can create severe environmental consequences if not properly accomplished.

Sulfide minerals contain metals that have combined only with sulfur. The lead ore galena (PbS), the copper ore chalcopyrite ($CuFeS_2$), and the zinc ore sphalerite (ZnS) are sulfide minerals that represent economically important metal ores. The mineral pyrite (FeS_2) is the most common sulfide mineral. Sometimes called iron sulfide, pyrite has a brass-yellow color and metallic luster that create an interesting physical appearance that earns this mineral its famous nickname *fool's gold*.

Carbonate minerals contain metals combined with carbon and oxygen. The mineral calcite ($CaCO_3$) resides within this group. Also known as calcium carbonate, calcite is a common constituent in sedimentary rocks, especially limestone. One of the more important applications of calcite is the production of cement.

Halide minerals include natural salts such as halite (sodium chloride [NaCl]), sylvite (potassium chloride [KCl]), and fluorite (calcium fluoride [CaF_2]). Sodium chloride (or salt) has been a very important mineral in the development of human civilization from ancient times up to the present. Potassium chloride has applications as a fertilizer, in the chemical industry, in medicine, and to produce decorative color effects on metal alloys such as brass and bronze. Depending on purity, fluorite, also known as fluorspar, has applications in the production of steel, manufacture of ceramics, and production of hydrofluoric (HF) acid.

AMERICAN HISTORY—CARVED IN STONE

Visitors to South Dakota can experience a breathtaking tribute to the concepts of liberty and freedom as manifested during the first 150 years of national history. Carved into the southeast face of a granite mountain called Mount Rushmore in the majestic Black Hills region of the state are the gigantic (59-foot [18 m]-tall) faces of four famous American presidents: George Washington (1732–99), Thomas Jefferson (1743–1826), Theodore Roosevelt (1858–1919), and Abraham Lincoln (1809–65). Officially called the Mount Rushmore National Memorial, the site is presently managed by the National Park Service of the U.S. Department of the Interior.

The Lakota Sioux Native Americans originally referred to this site as the Six Grandfathers. The mountain received its current name in 1885 after being viewed by the New York lawyer Charles E. Rushmore, who was traveling in the region as part of a mining claims and property titles expedition. In 1923, a

(continues)

The faces of the American presidents George Washington, Thomas Jefferson, Theodore Roosevelt, and Abraham Lincoln are carved in granite at the Mount Rushmore National Memorial in South Dakota. *(National Park Service)*

(continued)

state historian named Jonah Leroy "Doane" Robinson was considering how to attract more tourists to the Black Hills of South Dakota. One of his ideas was to use the granite pillar region of the Black Hills called The Needles as the site for several colossal carvings of western folk heroes. After further deliberation, the sculptor Gutzon Borglum (1867–1941) was approached, and he suggested carving large statues of four American presidents at a different location in the Black Hills known as Mount Rushmore. In Borglum's own words: "The purpose of the memorial is to communicate the founding, expansion, preservation, and unification of the United States with colossal statues of Washington, Jefferson, Lincoln, and Theodore Roosevelt." The sculptor favored the Mount Rushmore site because, facing southeast, it would receive good light for most of the working day. Mount Rushmore was also the highest peak in the vicinity (about 5,741 feet [1,750 m] above sea level), and its granite composition could successfully resist weathering and erosion. Since an exposed granite rock typically erodes about one inch (2.54 cm) per 10,000 years, the memorial's sculpted features could remain distinctive for thousands of years into the future.

Borglum started carving the mountain's rock on October 4, 1927. Assisted by a small army of workers (approximately 400), he used dynamite to remove rock from the mountain and to shape the broad features of each president's face. The sculptor completed the detailed carving of George Washington's face first. Consistent with the general theme of the memorial, Washington's face was officially dedicated in 1934 on Independence Day, July 4. The face of Thomas Jefferson was dedicated in 1936, followed by that of Lincoln in 1937. The dedication of Theodore Roosevelt's granite-carved face took place in 1939.

Borglum's original design for the stone memorial called for a colossal head-to-waist granite-carved image of each of the four presidents. When he died suddenly in March 1941, his son Lincoln Borglum attempted to continue the project. However, funding ran out seven months later, and the sculpting of Mount Rushmore ceased with the four presidential faces remaining "complete" as they stood at that moment.

In 1991, President George H. W. Bush officially dedicated the Mount Rushmore National Memorial. Today, this magnificent statement-in-granite about American history and liberty attracts millions of visitors annually to South Dakota's Black Hills.

NATURAL GEMSTONES

Geologists define a *natural gemstone* as a mineral, stone, or organic matter that can be cut and polished or otherwise treated for use as jewelry. Archaeologists and anthropologists point out that people have been adorning their bodies with special objects, including ornamental rocks and minerals, since prehistoric times.

Gems accompanied by precious metals, such as gold, silver, or platinum, are found in many types of personal jewelry, including finger rings, earrings, bracelets, and necklaces. Throughout history, reigning royal families in various societies, countries, and empires have often used crown jewelry (such as gem-encrusted golden crowns, orbs, scepters, rings, and swords) as tangible symbols of their authority and right to reign. The modern Imperial British Crown of State contains a total of 2,783 diamonds, 17 sapphires, 277 pearls, 11 emeralds, and 5 rubies. The collection of gems in this magnificent and priceless crown includes a very large, 317-carat diamond called the Second Star of Africa. The British monarch's royal scepter contains an enormous, 530-carat, pear-shaped diamond called the Great Star of Africa.

A gem is a gemstone that has been properly cut and polished. Gemologists are certified professionals who are trained to identify and evaluate gem materials. A lapidary is a person who cuts, polishes, or engraves precious stones. The preferred unit of mass for gems is called the *carat* (symbol: ct). This interesting unit of mass traces its origin back to antiquity and the tiny, uniform seeds of the carob bean. By current international agreement, a mass of one carat is equal to a mass of 4.4×10^{-4} lbm (200 milligrams [mg]). Gemologists also divide the carat up into 100 points, with each point representing a tiny mass of just 4.4×10^{-6} lbm (2 mg). (As discussed in chapter 6, jewelers and precious metal dealers use the term *karat* [symbol: kt] to describe the purity of various gold alloys.)

Within the fields of geology, mineralogy, and gemology, the following minerals are often regarded as precious stones: diamond, ruby, sapphire, emerald, topaz, and opal. All other gemstones, including amethyst, aquamarine, citrine, garnet, peridot, tourmaline, turquoise, and zircon, are generally considered to be semiprecious stones. The line between precious and semiprecious stones has changed over time. From ancient Greece through the middle of the 19th century, amethyst was widely regarded as a precious stone, but the discovery of large quantities of this mineral in Brazil in the 19th century reduced its scarcity and transformed its gemstone rating from precious to semiprecious. The ancient Greeks and Romans

Amethyst *(National Biological Information Infrastructure)*

considered amethyst to be a protection against intoxication. The gem remains popular today as the birthstone for February.

The chemical composition of a gemstone is of primary importance to both the mineralogist and the gemologist. Diamond is an amazingly hard allotrope of pure elemental carbon; corundum is a crystalline form of aluminum oxide (Al_2O_3) that yields either intensely red rubies or deep blue sapphires, depending on the impurities present in the gem material; and amethyst is a rich violet variety of quartz (silicon dioxide [SiO_2]). The mineral beryl consists of beryllium aluminum silicate ($Be_3Al_2[SiO_3]_6$). When tinted by impurities during formation, beryl yields a variety of interesting gemstones, including rich green emeralds and pale blue aquamarine stones. Gemstones are generally scarce and do not form extensive ore deposits in the normal mineralogical sense. When gemstones are present, they tend to be scattered sparsely throughout a large body of rock or have crystallized as small aggregates.

Two organic gemstones are briefly mentioned here. The pearl is formed within a mollusk such as an oyster when the living creature deposits a self-protective substance called nacre around an irritant that has lodged in its muscle tissue. Although pearl-bearing mollusks are found in both salt water and freshwater, saltwater pearls are the gem-quality materials used in fine jewelry. Natural pearls come in a variety of shapes, including round, egg, and pear; they also come in a variety of colors, including white, cream, light rose, bronze, blue, red, purple, and black.

Although not a mineral, amber has been used as a gemstone from prehistoric times up to the present. Amber is actually an amorphous mixture of hydrocarbons in the form of hard fossilized resin (or sap) from ancient trees. This organic gem material ranges from transparent to semitransparent and generally exhibits a characteristic light yellow to dark brown color. Orange, red, blue, and green color varieties of amber have also been found.

CHAPTER 6

Metals

Metals account for the great majority (more than 65 percent) of the elements found in the periodic table and represent about 25 percent of Earth's planetary mass. The rise of civilization beyond agriculture is essentially the story of the human mind understanding how to find and use metals. This chapter discusses the general properties of metals, introduces the metals of antiquity, and highlights several of the metals essential to humans' current global civilization.

GENERAL CHARACTERISTICS OF METALS

People usually identify metals as substances that have a luster or shine and are opaque. Sometimes an oxide layer forms on the metal's surface, giving it a dull appearance. An oxide layer occurs because the metal reacts with air. If the oxide layer is removed—by polishing or scratching—the metal's surface will once again shine.

Aluminum (Al) is a common example of this phenomenon. Although the most abundant metal in Earth's crust, aluminum is tightly bound in compounds, such as Al_2O_3, and is not found as a pure metal in nature. Freshly prepared, shiny aluminum metal reacts with oxygen in Earth's atmosphere and forms a thin, transparent layer of aluminum oxide. This hard oxide layer protects the metal and prevents further oxidation reactions from occurring. The metal iron (Fe) is not so "lucky." When exposed

CORROSION

Engineers use approximately 20 percent of all the iron and steel produced annually in the United States to replace metal items that have corroded. Corrosion takes place when a metal such as iron experiences chemical interactions with its environment and starts to disintegrate. In moist air, iron experiences oxidation, forming hydrated iron oxide ($Fe[OH]_3$), or rust.

Metals corrode when people use them in environments where the metals are chemically unstable. Only copper and certain precious metals such as gold, silver, and platinum are found in nature in their metallic state. All other metals, including iron, are bound up in mineral compounds, which must be processed in order to extract the pure metal. Once processed, these other metals are inherently unstable as pure metals and have a tendency to revert to their more stable mineral forms. Some metals, such as aluminum, form a protective passive film (typically oxide) on their surfaces.

There are many types of corrosion. Uniform corrosion and galvanic corrosion are described here. Engineers define *uniform* (or general) *corrosion* as the surface effect produced on a metal by most direct chemical attacks, such as by acid. The result is an overall uniform etching of the metal. Corrosion specialists first observe this phenomenon on a polished surface when they detect a general dulling of that surface. If they allow the corrosive attack to continue, the metal surface becomes rough and even frosted in appearance. Engineers use chemical-resistant protective coatings to help control this problem.

Galvanic corrosion takes place when dissimilar metals are in contact. Engineers define *galvanic corrosion* as an electrochemical reaction between two dissimilar metals in the presence of an electrolyte and an electron conductive path. This undesirable corrosive process is similar to how an electrochemical cell works in a battery. Engineers recognize the presence of this corrosive phenomenon when they observe the buildup of corrosion at the juncture of two dissimilar metals. When aluminum alloys are in contact with carbon steel, galvanic corrosion can take place, often resulting in the accelerated corrosion of the aluminum. One way engineers avoid galvanic corrosion is by using designs that electrically isolate the two dissimilar metals.

Understanding the complexities of corrosion and preventing the deterioration of metals used in engineered systems is an important part of materials science.

to air, a shiny piece of iron reacts extensively with atmospheric oxygen to form iron oxide (Fe_2O_3), more commonly known as rust. Other metals, such as gold (Au), resist corrosion and maintain their luster.

Metals are generally solid at room temperature (nominally 68°F [20°C {293 K}]), are dense, have distinctive crystalline structures, and are good conductors of heat and electricity, but there are exceptions to these general characteristics. Lithium (Li), a silver-white colored alkaline metal, is the least dense solid element and the lightest metal. With a melting point of –37.9°F (–38.8°C [234.3 K]), mercury (Hg) is a liquid under room temperature conditions. The metal gallium (Ga) has a melting point of 85.6°F (29.8°C [302.9 K]), which means this metal is a liquid at just above room temperature, so a small piece of it would melt in the palm of a person's hand. Similarly, cesium (Cs) is another metal that transitions to the liquid state at just above room temperature conditions. It is an alkali metal with a melting point of 83.2°F (28.4°C [301.6 K]). Bromine (Br), with a melting point of 19.0°F (–7.2°C [266 K]), is the only nonmetallic element that is liquid at room temperature.

Chemists and physicists correlate the properties of the elements portrayed in the periodic table with their electron configurations. Since in a neutral atom the number of electrons equals the number of protons, scientists arrange the elements in order of their increasing atomic number (Z). The modern periodic table has seven horizontal rows called periods and 18 vertical columns called groups. The properties of the elements in a particular row vary across it, providing the concept of periodicity. Scientists refer to the elements contained in a particular vertical column as a group, or family. (See appendix on page 164.)

Metals are those elements that have few electrons in the outermost orbital shell. Scientists have discovered that elements with only one electron in the outermost shell typically have the highest conductivity (electrical and thermal). Silver (Ag), copper (Cu), and gold (Au) are metals that exhibit high electrical and thermal conductivity. An inspection of the periodic table indicates that metals occur on the left side and in the middle portion of the table. Most metals are malleable, a characteristic that allows artisans to hammer a particular metal into very thin, flat sheets. Finally, metals are generally ductile, the property that allows them to be drawn into long, narrow wires.

Chemists and physicists find it useful to describe families of elements using the following terms: alkali metals, alkaline earth metals, transition metals, basic (or other) metals, metalloids, nonmetals, halogens, and

GOLD'S ELECTRON CONFIGURATION

Gold is an excellent conductor of both electricity and heat. Since this metal does not corrode and remains very stable, engineers favor its use in the efficient and reliable transport of the tiny amounts of current within modern solid-state electronic devices, microprocessors, and computer memory chips. Scientists use an electron configuration table, like the one shown below, to describe how many electron energy levels are in a particular elemental atom and how these electrons are arranged within each energy level. The electron configuration table for a gold atom is as follows:

$1s^2$	Energy level one (lowest)
$2s^2\ 2p^6$	Energy level two
$3s^2\ 3p^6\ 3d^{10}$	Energy level three
$4s^2\ 4p^6\ 4d^{10}\ 4f^{14}$	Energy level four
$5s^2\ 5p^6\ 5d^{10}$	Energy level five
$6s^1$	Energy level six (highest)

Scientists use a relatively simple code to explain how a neutral gold atom's 79 electrons are arranged. The first number (such as 1, 2, 3 . . .) represents the energy level. Through the table shown above, scientists tell other scientists that a gold atom's electrons occupy six energy levels. The lowercase letter (here s, p, d, or f) represents the subshell to which an electron belongs. The number of available subshells increases as the energy level increases. For example, subshell "s" can hold only two electrons, while subshell "p" can hold a maximum of six electrons. The superscripted number (at the right) expresses the number of electrons in each particular subshell. The notation "$6s^1$" tells scientists that gold has only one electron in the highest (or sixth) energy level.

noble gases. A few characteristics of each these families are presented here. For a more comprehensive discussion, the reader is referred to any contemporary college-level chemistry textbook.

The alkali metals are the family of elements found in group 1 (column 1) of the periodic table. They include lithium (Li), sodium (Na), potassium (K), rubidium (Rb), cesium (Cs), and francium (Fr). These metals react vigorously with water to produce hydrogen gas. For safety, these metals are generally stored under mineral oil.

The alkaline earth metals are the family of elements found in group 2 (column 2) of the periodic table. They include beryllium (Be), magnesium (Mg), calcium (Ca), strontium (Sr), barium (Ba), and radium (Ra). These elements are chemically reactive, but less so than the alkali metals. The soft, silver-colored metals react with water (though not as aggressively as the alkali metals) to form strongly alkaline hydroxides such as calcium hydroxide $Ca(OH)_2$. Beryllium is the somewhat unusual member of this elemental family, since it is rather hard, is steel gray in color, and does not react with water.

The transition metals are the metallic elements located in the central portion of the periodic table and include the elements found in groups (columns) 3 to 12. (See appendix on page 164.) Chemists describe the transition metals as the group of elements characterized by the filling of an inner (*d* electron) orbital as the atomic number increases.

The border between metals and nonmetals on the right side of the periodic table is somewhat chaotic, resulting in steplike partitions of the elements. One such partition occurs when scientists call the metals found in group (column) 12 (that is, zinc, cadmium, and mercury) *posttransition metals.* Not all chemists use this particular nomenclature and, for simplicity, neither does this book.

Transition elements generally have high density, high melting point, and (some) magnetic properties. Containing about two-thirds of all the elements in the periodic table, some of the more familiar transition metals are (in order of ascending atomic number) titanium (Ti), chromium (Cr), iron (Fe), cobalt (Co), nickel (Ni), copper (Cu), silver (Ag), tungsten (W), gold (Au), and mercury (Hg).

Since the transition metals include the majority of the elements, scientists sometimes find it useful to identify smaller groupings or categories of transition metals. They regard the lanthanoids (formerly lanthanides) and actinoids (formerly actinides) as classes of metals within the large family of transition metals.

Chemists and metallurgists sometimes group three similar transition metals into smaller groups, called *metal triads.* The iron triad contains iron (Fe), cobalt (Co), and nickel (Ni); the palladium triad consists of ruthenium (Ru), rhodium (Rh), and palladium (Pd); and the platinum triad contains osmium (Os), iridium (Ir) and platinum (Pt).

The lanthanoid (formerly lanthanide) series contains 15 elements that start with lanthanum (La) and continue through lutetium (Lu). All the elements in the lanthanoid series closely resemble the element lanthanum. Scientists often collectively refer to the 15 elements in the lanthanoid series along with the element yttrium (Y) as the *rare earth elements.*

Generally, members of the lanthanoid series are silvery solids that tarnish easily. These relatively soft metals have high melting points and react to form many compounds. Engineers use these elements in many important modern technology applications, including intense magnets, lasers, special lamps, magnetic refrigeration devices, advanced batteries, and hydrogen storage. Small, low-mass, high-strength rare earth element magnets have allowed engineers to miniaturize numerous electrical and electronic components, such as multigigabyte portable disk drives. The term *rare earth element* is actually a historic misnomer. According to scientists at the U.S. Geological Survey (USGS), the more abundant rare earth elements, such as cerium (Ce), are similar in crustal concentrations to such commonly used industrial metals as chromium, nickel, copper, and zinc.

The collection of new, human-made, superheavy elements (SHEs) starts with element number 104 (called rutherfordium [Rf]) and continues through element 118 (called ununoctium [Uuo]). These superheavy elements appear in row (period) 7 of the periodic table after the element actinium (Ac). (See appendix.) Fleeting amounts (one to several atoms) of all these elements, have been created in complex laboratory experiments. The elements from 104 to 118 have extremely short half-lives, typically milliseconds to seconds, making any type of physical property analysis extremely difficult.

Chemists and metallurgists often classify the following seven elements as *basic metals* (or *other metals*): aluminum (Al), gallium (Ga), indium (In), tin (Sn), thallium (Th), lead (Pb), and bismuth (Bi). While the metals listed in this category are ductile and malleable, they do not exhibit the same oxidation state properties as the transition elements. Basic metals reside in groups (columns) 13, 14, and 15 of the periodic table—forming a staircaselike border between the transition metals and the metalloids.

Scientists define *metalloids* as elements that exhibit properties intermediate between the transition metals and the nonmetals. The elements boron (B), silicon (Si), germanium (Ge), arsenic (As), antimony (Sb), tellurium (Te), and polonium (Po) are considered metalloids. These elements form a distinctive steplike pattern at the right side of the periodic table, involving groups (columns) 13, 14, 15, and 16.

Groups (columns) 13, 14, and 15 of the periodic table contain some metals, some metalloids, and some nonmetals. The transition from metal to nonmetal is gradual. Chemists often group the following elements into the overall category of nonmetals: carbon (C), nitrogen (N), oxygen (O), phosphorous (P), sulfur (S), and selenium (Se). Although hydrogen (H) is a member of group 1 of the periodic table, the cosmically ubiquitous element often appears in the list of nonmetal elements.

Chemists identify the halogens as the series of nonmetal elements from group (column) 17 of the periodic table. The halogens include fluorine (F), chlorine (Cl), bromine (Br), iodine (I), and astatine (As). Fluorine, a pale yellow-colored gas, is the most chemically reactive of all the elements and does not occur free in nature. Chlorine is a greenish-yellow gas that is reactive and toxic. Bromine is a caustic, foul-smelling, reddish-brown, nonmetallic element that is liquid at room temperature.

Finally, the noble gases are the inert gaseous elements found in group (column) 18 of the periodic table. These elements include helium (He), neon (Ne), argon (Ar), krypton (Kr), xenon (Xe), and radon (Rn). The noble gases do not readily enter into chemical combination with other elements. Human-produced nuclear reactions result in the production of certain radioactive isotopes of krypton and xenon, while radon is a naturally occurring source of radioactivity.

THE METALS OF ANTIQUITY

Materials scientists often regard process metallurgy as one of the oldest applied sciences. The first important step that led to materials science and technology was the discovery and use of fire in prehistoric times. As part of the Neolithic Revolution, craftsmen learned how to make fire-hardened bricks and ceramic pottery. The kilns used by brick makers and potters would play a major, though unexpected, role in the ascendancy of humanity. The *kiln* is a specially designed oven capable of providing a well-controlled internal temperature over a sufficient volume to dry, heat, or melt the material being processed. These discoveries proved of immense importance in the development of the first cities and in the preparation, storage, and transport of food, but it was the use of metals, especially copper, tin, and iron, that placed early human civilizations on a technology-based trajectory that continues to this very day. Without metals, human beings would have remained locked in resource-limited, localized, agrarian-based societies.

The metals of antiquity were gold, copper, tin, silver, lead, mercury, and iron. These metals are individually discussed in subsequent sections in the approximate order of their use in historic times. The exact circumstances surrounding each discovery remain clouded in history. Supported by artifacts found in tombs, archaeologists suggest the early periods in human history when the use of each metal first occurred.

The use of metals and other materials (such as carbon [C] and sulfur [S]) supported the rise of advanced civilizations all around the Mediterranean Sea, with the Roman Empire serving as an enormous political stimulus and evolutionary endpoint. When the Roman Empire collapsed, western Europe fragmented into many small political entities. Progress in materials science continued there, but at a much more fractionated and sporadic pace—until the 17th century and the start of the scientific revolution.

Gold

In all likelihood, Stone Age human beings used native gold (found as nuggets) in primitive forms of jewelry. This tarnish-free metal has a captivating color that continues to fascinate humans to this day. Sometime around 6000 B.C.E., early artisans (later called goldsmiths) discovered the malleability of gold and began hammering nuggets into very thin sheets of metal and wires for use in jewelry and ornaments, including crowns. However, because gold is so ductile and soft, the metal did not find more practical uses in ancient societies as tools or weapons.

Gold has been an attractive, highly valued metal since antiquity. The term *gold fever* describes the compelling desire of people throughout the ages to seek and hoard gold. Gold has an atomic number of 79 and an atomic weight of 196.967. This lustrous, soft yellow metal has a density of 1,203.8 lbm/ft^3 (19.282 g/cm^3) at room temperature and a melting point of 1,947.52°F (1,064.18°C [1,337.33 K]). Gold is found as a free metal in nature; it is also found in conjunction with quartz (SiO_2), calcite ($CaCO_3$), lead, tellurium, zinc, and copper. Metallurgists note that gold is the most malleable and ductile of all metals. Skilled goldsmiths can hammer one troy ounce (31.1 g) of gold into a very

Gold leaf on quartz *(USGS/BOM)*

thin sheet about 300 ft^2 (27.9 m^2) in total area. Gold does not tarnish. Instead, the metal resists corrosion and maintains its luster. Gold is also a good conductor of both heat and electricity. At 68°F (20°C [293 K]), this transition metal has a thermal conductivity of 183.7 Btu/(h-ft-°F) (318 J/[s-m-K]) and an electrical conductivity of 48.8×10^6 S/m.

Throughout human history, artisans have used gold and gold leaf to lavishly decorate temples, tombs, and sacred objects. *Gold leaf* is metallic gold that has been beaten into an incredibly thin foil. It is a gilding material used to decorate statues, carvings, picture frames, the roof tops of sacred buildings, and similar ornate objects.

The major use of gold today is in jewelry. Since pure gold is generally too soft to withstand the physical rigors that jewelry sometimes experiences in daily wear, goldsmiths often alloy gold with other metals such as nickel, silver, or copper, but gold alloys have a lower intrinsic value than pure gold, so metallurgists created a system to define the pureness of gold. Historically, the *karat* (symbol K or Kt) defined the purity of gold alloys. One karat represents gold that is 1/24 purity by mass—that is, one part gold and 23 parts other metal(s) by mass. Pure gold is 24-karat gold, while an alloy that contains 75 percent gold by mass is called 18-karat gold.

Modern metal traders and bullion investors prefer to use another system, called the *millesimal fineness system.* In this system, the purity of gold, silver, and platinum alloys is expressed as parts per thousand of the pure metal in the alloy. *Bullion* is gold, platinum, or silver formed into bars, ingots, or special (noncirculating) coins that have been stamped with the metal's purity, an inscription called the fineness hallmark. For gold, a fineness value of 999 is referred to as "fine gold" and corresponds to 24-karat gold. Similarly, a fineness value of 750 represents a gold alloy equivalent to 18-karat gold.

In years past, governments often supported their national monetary systems with large reserves of gold bullion, but the size of today's world economy (estimated to be in excess of $65 trillion in 2009) makes that economic practice impractical. The United States Mint reported in 2009 that the U.S. Bullion Depository at Fort Knox, Kentucky, presently holds 147.3 million troy ounces (about 4.58×10^6 kg) of gold. Assuming gold has a value of $1,000 per troy ounce, the resulting market value of all the gold in Fort Knox would then be about $147 billion—well below the value of all the American money in circulation and government-backed financial instruments. The weight of precious metals such as gold and gemstones is typically expressed in troy ounces, where one troy ounce equals 31.103

grams. The troy weight system originated in medieval times, and its name derives from the French town of Troyes.

Other uses of gold include the production of bullion coins (for investment), dentistry, electronics, research, medical therapy, aerospace applications, and awards. Aerospace engineers use very thin layers of gold on some of a spacecraft's outer surfaces to help maintain the space vehicle's internal temperatures within design limits. The significance of gold is very deeply embedded in human culture. The crowns of kings and emperors are usually made of gold. First-place medals, trophies, and awards for excellence are generally made of gold. In contrast, second-place awards are usually honored by the metal silver, and third place by the copper-tin alloy bronze.

Copper

Copper (Cu) is a soft, reddish-orange metal that can assume a bright metallic luster. Historians generally consider copper to be the most important metal in the rise of human civilization because the first metal tools, implements, and weapons were fashioned out of copper.

Around 5000 to 4500 B.C.E., artisans began hammering native copper (also found in the form of nuggets) into sheets of metal, but once hammered (cold-worked), copper became brittle. By accident or clever experimentation, some early artisan discovered how to heat treat (anneal) the hammered copper, and the metal became much more useful for tools and weapons. After the relationship between fire and the processing of metals was discovered, smelting soon emerged. Materials scientists define *smelting* as the general practice of using heat and possibly a chemical reducing agent to extract a metal from its ore. A blast furnace usually facilitates this form of extractive metallurgy.

Archaeological evidence suggests that the early peoples of the Nile River Valley learned how to coax copper out of the green-colored mineral malachite (copper carbonate, or $CuCO_3Cu[OH]_2$) by heating it in a certain way. One possibility is that an ancient potter accidentally placed a piece of malachite into a high-temperature (about 2,192°F [1,200°C {1,473 K}]) kiln and then discovered nodules of copper metal at the bottom of the kiln while removing the ceramic articles after firing. The actual discovery remains clouded in antiquity, but soon royal persons around the eastern regions of the Mediterranean Sea, including the rulers of Egypt, were wearing copper armor and wielding copper swords in combat.

Artisans in the Middle East (most notably the Sumerians) soon recognized that smelted copper was rarely pure. Different ores produced different results, including a blend of copper and tin (Sn)—about 89 percent Cu and 11 percent Sn by weight—that proved to be a much better metal. They discovered the alloy bronze. The civilization-transforming event triggered a resource stampede around the eastern Mediterranean Sea for metal-bearing ores.

The Bronze Age occurred in the ancient Middle East starting about 3500 B.C.E. and lasted until about 1200 B.C.E. During this period, metalworking artisans began to regularly use copper and bronze to make weapons and tools. As the demand for copper and bronze products grew, peoples of the early civilizations, such as the Minoans on Crete, expanded their search for additional sources of copper and tin. Bronze Age mercantile activities encouraged exploration and trade all around the Mediterranean Sea. The Minoans sponsored settlements on Cyprus and developed copper mines there. Cyprus is the third largest island in the Mediterranean Sea. During the Roman Empire, the Latin phrase *aes Cyprium* meant "metal of Cyprus" and was eventually shortened to *cuprum*—the Latin name for copper and also the source of the chemical element symbol Cu.

Copper has an atomic number of 29 and an atomic weight of 63.546. This transition metal has a density of 557.7 lbm/ft^3 (8.933 g/cm^3) at room temperature and a melting point of 1,984.32°F (1,084.62°C [1,357.77 K]). At 68°F (20°C [293 K]), copper has a specific heat capacity of 0.091 Btu/lbm-°F (0.38 kJ/kg-K), a thermal conductivity 231.7 Btu/(h-ft-°F) (401 J/[s-m-°C]), and an electrical conductivity of 60.7×10^6 S/m.

Copper is found in native form as well as in such minerals as malachite ($Cu_2CO_3[OH]_2$) and cuprite (Cu_2O). Copper is second only to silver in electrical conductivity, so engineers use this soft (3.0 on the Mohs scale), malleable, ductile metal in large quantities in the electrical industry, especially in the form of wires of all diameters. Copper is also used in coins, water pipes, and jewelry.

Since pure copper is generally too soft for most applications, people learned to mix copper with other metals to form alloys with superior properties. The mixture of copper and tin became bronze, an alloy that gave its name to an entire period of human development, the Bronze Age. The Romans became the first to make extensive use of brass, a mixture of copper and zinc. The horn and trumpet were the earliest brass musical instruments. Today, the brass section of a modern symphonic orchestra resounds with blazing fanfares.

Copper is also used by artists and architects to achieve special effects. Copper interacts with moist air to form a surface coating called *patina.* This patina protects the metal from further interaction or corrosion. The

Six U.S. Air Force F-16 Fighting Falcon jet aircraft fly in formation over the Statue of Liberty, a world-renowned icon of liberty and freedom. Patination of the famous statue's copper skin provides its rich green color. Patinas on metals are caused by the corrosive action of chemicals and environmental conditions. Some metals, such as copper, develop a patina that is often extremely attractive. *(U.S. Air Force)*

green patina that forms naturally on copper consists of copper carbonate ($CuCO_3$) or copper chloride ($CuCl_2$). People sometimes refer to the green patina as verdigris.

The Statue of Liberty on Liberty Island in New York Harbor has a rich green patina. The French sculptor Frédéric Bartholdi (1834–1904) created the statue's copper skin by hammering pieces of copper (on the reverse

ALLOYS AND SUPERALLOYS

Materials scientists define an *alloy* as a solid solution (compound) or homogeneous mixture of two or more elements, at least one of which is an elemental metal. An alloy is a metallic substance with different characteristics and properties than those of the components. The first alloy used by human beings was bronze, a mixture of copper and tin. Brass is a mixture of copper and zinc. Steel is an alloy of iron, carbon, and small amounts of other elements such as chromium, molybdenum, or nickel. The common soft (low-temperature) solder used by electricians is a tin-lead alloy (typically about 50 percent by weight of each element). Pewter is a ductile, silver-white alloy consisting mostly of tin that is usually mixed with either lead or antimony. Antimony makes the pewter hard and white; lead gives pewter a bluish tint and makes the alloy more ductile. Artisans also add copper, bismuth, or zinc to produce pewter with different properties.

Metallurgists define a *superalloy* as an alloy that resists corrosion and oxidation at high temperatures. Such materials maintain dimensional stability and mechanical strength at elevated temperatures. Materials scientists typically use nickel or cobalt as the base alloying element in creating a superalloy. They then blend varying amounts of other elements—such as aluminum, carbon, titanium, molybdenum, chromium, tungsten, or zirconium—to form substances that can dependably perform when exposed to the extreme environments often encountered by components in modern aerospace, defense, and industrial systems. The turbine blades of a modern jet aircraft are just one example.

side) until each piece was properly shaped and only 3/32 of an inch (0.238 cm) thick—about the thickness of two American pennies. The 151-foot (46 m)-tall statue was erected over an iron and steel skeleton on a large concrete pedestal and originally dedicated on October 28, 1886. The statue contains approximately 62,000 lbm (28,120 kg) of copper. The National Park Service (NPS) of the U.S. Department of the Interior has responsibility for maintaining this iconic monument to freedom.

Tin

Tin (Sn) has an atomic number of 50 and an atomic weight of 118.710. Known since antiquity, tin is generally a silver-white metal that has a highly crystalline structure with a density of 454.9 lbm/ft^3 (7.287 g/cm^3) at room temperature. The element has two allotropic forms at normal atmospheric pressure. Gray (or alpha) tin changes its crystalline structure and becomes white (or beta) tin when the metal's temperature reaches or exceeds 55.8°F (13.2°C). There are few applications of gray tin, so the remaining comments pertain to so-called white tin. Tin is primarily extracted from the mineral cassiterite (SnO_2).

Because tin resists corrosion, metallurgists often use it as a protective coating on other metals. Tin cans are a familiar example of this application. A tin can is actually made from sheets of steel that have a thin layer of tin applied on both sides to prevent rusting. Materials scientists use tin to make many useful alloys, including bronze (copper and tin) and pewter (tin and lead). Tin is malleable and somewhat ductile. It has a melting point of 449.47°F (231.93°C [505.08 K]). The compound stannous fluoride (SnF_2) is found in some types of toothpaste.

Silver

Silver (Ag) is found freely in nature and also in ores commingled with gold. Early metalworking artisans (about 4000 B.C.E.) began fashioning silver into jewelry and ornaments. Soft and ductile, like gold, silver did not lend itself to practical use in tools, implements, or weapons. It did, however, begin to represent the accumulation of wealth. Early peoples used the term *electrum* to identify the naturally occurring alloy of gold and silver.

THE INVENTION OF MONEY

Archaeologists suggest that human beings used the barter system at least 100,000 years ago. As trading activities increased within and between the early civilizations emerging in the ancient Middle East and around the Mediterranean Sea region, the need for a more efficient means of transferring goods and storing wealth also arose. In about 3000 B.C.E., the people of Mesopotamia began using a unit of weight called the shekel to identify a specific mass of barley (estimated to be about 0.023 lbm [10 grams]). The shekel became a generally accepted weight of convenience for trading other commodities, such as copper, bronze, and silver. In about 650 B.C.E., the Lydians, an Iron Age people whose kingdom was located in the western portion of modern Turkey, invented money when they introduced officially stamped pieces of gold and silver that contained specific quantities of these precious metals. Money has been the lubricant of global economic activities ever since.

Silver is a shiny lustrous metal that people have used since antiquity in jewelry, ornamental objects, and coins. Silver has an atomic number of 47 and an atomic weight of 107.868. This transition metal has a density of 655.6 lbm/ft^3 (10.501 g/cm^3) at room temperature and a melting point of 1,763.20°F (961.78°C [1,234.93 K]). Pure silver has the highest value of thermal conductivity and electrical conductivity of all metals. At 68°F (20°C [293 K]), silver has a specific heat capacity of 0.056 Btu/lbm-°F (0.235 kJ/kg-K), a thermal conductivity of 247.9 Btu/(h-ft-°F) (429 J/[s-m-°C]), and an electrical conductivity of 63.0×10^6 S/m. Although silver conducts electricity better than copper, economic factors encourage engineers to use copper in the overall generation and distribution of electric power.

The element is found as a native metal in the ore argentite (Ag_2S) and in conjunction with ores containing gold, copper, and lead. Pure silver is a shiny, ductile, malleable metal that artisans can beat into thin foil and draw out into long strands and wires. Unlike gold, silver tarnishes when exposed to environmental sulfur compounds that may be present in air or water. When silver tarnishes, the metal forms an unsightly black sulfide layer on its surface. As a pure metal, silver is the best-known reflector of visible light. In the past, people used silver for mirrors, even though the metal would quickly tarnish.

Silver is a soft metal (about 2.5 on the Mohs hardness scale), so artisans would often add other metals, such as copper, to produce silver alloys that were more suitable for use as utensils, instruments, and durable jewelry. *Sterling silver* is a common silver alloy. It consists of 92.5 percent silver, with the remainder of the mass being that of another metal, usually copper. Starting in antiquity, silver was often used as portable wealth and money. Today, people collect bullion coins and ingots hallmarked as containing "999 fine" silver in an effort to guard against inflation and weather uncertain economic conditions.

Silver remains an important metal with many industrial applications. Up until the age of digital photography, about one-third of the silver consumed each year in the United States went into various photographic materials and processes, primarily as silver bromide (AgBr) and silver nitrate ($AgNO_3$), which are light-sensitive compounds. Silver is also used in the manufacture of high-reliability electrical contacts and circuit boards, as an alloy in brazing and solder, in dental amalgams, and as a catalyst in the chemical industry. Finally, silver-zinc and silver-cadmium combinations are part of high-capacity batteries.

Lead

Another shiny mineral, called galena (lead sulfide [PbS]), became the source of the metal lead (Pb) in about 3000 B.C.E. The discovery of lead could have taken place when a piece of galena used for jewelry fell into an ancient camp fire and yielded beads of the metal. Because of its ductility and low melting point, Roman engineers became especially fond of fashioning lead into pipes, liquid containers, and conduits.

Lead has an atomic number of 82 and an atomic weight of 207.2. Known since ancient times, it is a very soft, blue-gray metallic element that has a density of 708.1 lbm/ft³ (11.342 g/cm³) at room temperature and at one atmosphere pressure. The melting point of lead is 621.43°F (327.46°C [600.61 K]). Although some lead is found as metal in nature,

GUTENBERG'S LEAD ALLOY REVOLUTIONIZES THE WORLD

In the middle of the 15th century, the German goldsmith and printer Johannes Gutenberg (ca.1400–68) assembled a number of innovative ideas into the world's first successful mechanical printing press with movable metal type. His invention revolutionized the mass production of books in Europe and started the wave of information technology improvements that continues to the present day. Central to Gutenberg's great contribution was his development of *type metal,* a special alloy of lead, tin, and antimony. This lead alloy possessed all the physical attributes needed to make a mechanical printing press with movable metal type possible and practical. The dominant metal was lead—abundant, inexpensive, and easily molded into individual letters and symbols, but lacking the necessary hardness and wear-resistance to consistently transfer sharp impressions to paper. Through the patient addition of different amounts of tin and antimony, Gutenberg eventually developed the right combination of metals—an alloy referred to as *type metal.*

Gutenberg's mechanical printing technology spread quickly throughout Europe, encouraging the dissemination of knowledge and the exchange of new ideas. Relatively inexpensive books became the catalyst of both the Renaissance and the scientific revolution. Science historians treat Gutenberg's invention of the mechanical printing press with movable metal type as one of humankind's most important technology achievements.

most is contained in lustrous mineral compounds, such as galena. The ancient Romans obtained lead by roasting galena in hot air. They used lead to make water pipes, some of which still carry water. The modern words *plumber* and *plumbing* come from the Latin word for lead, *plumbum*—which is also the source of the element's chemical symbol, Pb.

Contemporary uses of lead include lead-acid automobile batteries, ballast and weights, special glass crystals, ammunition, and shielding against X-ray and gamma ray radiation. Lead is toxic, so earlier uses in paints, insecticides, and gasoline have largely been discontinued in the United States. Engineers and technicians use the lead-tin alloy known as *solder* to join electrical components, pipes, and other metallic items. The lead-tin-antimony alloy called *type metal* composed the movable metal type used in printing presses.

Mercury

In about 1600 B.C.E., the metal mercury (Hg), a liquid at room temperature, also became known to ancient peoples around the Mediterranean Sea. This metal soon gained wide use because of its ability to dissolve gold and silver. Gold- and silver-bearing ores were crushed and treated with mercury. As the liquid mercury formed a solid amalgam, it extracted the precious metals from the host ore. An *amalgam* is an alloy containing mercury. Heat treatment of this particular amalgam would drive off the mercury, leaving behind the sought-after gold and silver.

Mercury (Hg) has an atomic number of 80 and an atomic weight of 200.59. Sometimes called *quicksilver,* mercury is usually not found free in nature and is the only common metal that is liquid at room temperature. This silver-white liquid metal has a density of 845.7 lbm/ft^3 (13.546 g/cm^3) at room temperature and breaks up into tiny drops or beads when spilled. Mercury is generally obtained from the mineral cinnabar (HgS). The liquid metal has a melting point of –37.89°F (–38.83°C [234.32 K]) and readily forms alloys (amalgams) with gold, silver, zinc, and cadmium. Mercury is toxic and a major environmental pollutant. The liquid metal is used in scientific instruments such as thermometers and barometers, fluorescent lamps, and mercury vapor streetlights. The compound mercuric oxide (HgO) is used in the manufacture of mercury batteries.

Iron

The last metal of antiquity dealt with here is iron (Fe). Originally, iron was available only in very small quantities because it came from meteorites. Iron is not found in the free state as a metal on Earth's surface. About 1200

B.C.E., some civilizations in the ancient Middle East entered the Iron Age when their metalworking artisans discovered various iron smelting techniques. Civilizations such as the Hittite Empire of Asia Minor enjoyed a military advantage as their artisans began to produce iron weapons and tools. Later during this period, the very best weapons and tools were those made of steel, an alloy of iron and varying amounts of carbon. Over the next six centuries, the use of iron and steel tools and weapons spread rapidly to Greece, Rome, and other parts of Europe.

This metal is a transition element that makes up about 5 percent of Earth's crust and a good portion of its deep interior. Pure iron is a dark, silvery-gray metal, but it is a very reactive element and oxidizes (forms rust) quite easily. Iron has an atomic number of 26 and an atomic weight of 55.845. This important metal has a density of 491.6 lbm/ft^3 (7.874 g/cm^3) at room temperature and a melting point of 2,800°F (1,538°C [1,811 K]). At 68°F (20°C [293 K]), iron has a specific heat capacity of 0.105 Btu/lbm-°F (0.44 kJ/kg-°C), a thermal conductivity of 46.5 Btu/(h-ft-°F) (80.4 J/[s-m-°C]), and an electrical conductivity of 0.112×10^6 S/m.

Iron is usually not found free in nature on Earth's surface. Rather, the metal is found in iron-bearing ores such as magnetite (Fe_3O_4) and hematite (Fe_2O_3). Any iron metal found on or near Earth's surface is the result of meteorite impacts. Approximately 98 percent of the iron ore annually mined throughout the world is used to make steel, one of the most important solid materials ever produced by human beings. Steel is an alloy of iron. It consists mostly of that metal and varying amounts of carbon (typically, 0.2 percent to 2.0 percent by weight) and small amounts of other elements such as chromium, manganese, nickel, or tungsten.

Making various types of steel is as much a metallurgical art as it is a science, with complex chemical reactions and interesting physical transitions influencing the final product. Carbon steel alloys that contain small amounts of chromium are more durable and represent a more rust resistant metal called *stainless steel.* From skyscrapers to suspension bridges, industrial machinery to ships, and paper clips to automobile frames, steel in all its various grades forms the metallic backbone of modern civilization.

Iron is the world's most extensively used metal. It is also the most magnetic of the three naturally magnetic elements, the other two being cobalt and nickel. Iron plays an important role in the health of living systems on Earth. Human blood cells rich in iron carry oxygen from the lungs to all parts of the body. The body needs iron to produce the necessary amount of hemoglobin, the respiratory molecule found in red blood cells. An iron deficiency results in the medical condition called anemia.

Metallurgists use heat to process iron ore and create good-quality steels. They must remove impurities, carefully control the carbon content, and blend in any other alloying materials at just the right time. Some of the terms commonly associated with the manufacture of steel are pig iron, wrought iron, carbon steel, and alloy steels. Materials scientists define *pig iron* as iron containing from about 3.5 percent to 4.5 percent by weight carbon and other contaminants such as silicon, phosphorous, and sulfur. *Wrought iron* is almost pure iron, containing less than 0.25 percent carbon by weight. It is tough though malleable and ductile. *Plain carbon steels* are iron alloys with less than 2 percent by weight carbon, accompanied by up to 0.9 percent by weight manganese and small amounts of sulfur, silicon, or phosphorous. Finally, *alloy steels* contain iron with varying amounts of carbon and other metals such as chromium, nickel, or tungsten. These alloys are custom-designed and blended for high strength and toughness. They are typically used in specific applications, such as skyscraper or bridge construction.

METALS AND MODERN CIVILIZATION

The metals of antiquity and the emerging science of metallurgy served as the building blocks of today's technology-enabled civilization. Industrialized countries such as the United States now use large quantities of metals. The pace of technical discoveries during the 20th century caused pivotal changes in the way minerals are produced and consumed. In steelmaking, for example, the basic oxygen and electric furnaces replaced the Bessemer and open-hearth furnaces. In 1855, the British engineer Sir Henry Bessemer (1813–98) patented an inexpensive process for the mass production of steel. Called the Bessemer process, his technique blew oxygen through molten pig iron to drive off impurities and form steel with the right carbon content. Because of rapid advances in aerospace, communications, construction, defense, and transportation technologies, lightweight, lowdensity metals, such as titanium and aluminum are very important in modern civilization.

Titanium

Titanium (Ti) is a lustrous, silver-white metal that has an atomic number of 22 and an atomic weight of 47.90. Considered a transition metal, titanium has a hexagonal crystalline structure. Some of its more important

FAMOUS BLACKBIRD WITH TITANIUM WINGS

The United States Air Force (USAF) developed and deployed the Lockheed SR-71 strategic reconnaissance aircraft in the 1960s. Nicknamed the Blackbird, the SR-71 retains the title of the world's fastest and highest-flying production aircraft. This spectacular jet aircraft could fly at more than 2,125 mph (3,400 km/h) and at altitudes of more than 85,000 feet (25,910 m).

Designing an aircraft for sustained flight at more than three times the speed of sound (Mach 3+) presented the engineers at Lockheed's famous Skunk Works® in Southern California with enormous technical challenges. Their success was due in no small measure to the use of the metal titanium. In a marvel of aerospace engineering and materials science, the engineers fabricated the airframe of the SR-71 almost entirely of titanium and titanium alloys. This enabled the aircraft to survive sustained flight at supersonic speeds. The outer surface of the titanium structure was coated with a special black paint, giving the SR-71 its famous nickname. The SR-71 served the defense and aerospace research needs of the United States from 1964 to 1998.

An SR-71B from NASA's Dryden Flight Research Center in California slices across the sky above the snowy southern Sierra Nevada Mountains. Originally used by the military as a reconnaissance platform, this supersonic aircraft's high-temperature air frame was made of titanium. ***(NASA)***

After retiring them from operational defense missions, the USAF provided NASA with several SR-71 aircraft for high-speed aircraft research projects in the 1990s. Because they could cruise at Mach 3 for more than one hour, NASA engineers employed these aircraft as special research platforms. NASA based the aircraft at the Dryden Flight Research Center in Edwards, California.

physical properties at room temperature are a density of 280.9 lbm/ft^3 (4.50 g/cm^3), specific heat of 0.124 Btu/lbm-°F (0.52 kJ/kg-°C), thermal conductivity of 12.65 Btu/(h-ft-°F) (21.9 J/[s-m-°C]), and electrical conductivity of 2.6×10^6 S/m. At one atmosphere pressure, this metal has a melting point of 3,034°F (1,668°C [1,941 K]).

Titanium was discovered in 1791 by the British clergyman and scientist William Gregor (1761–1817), then rediscovered several years later in 1795 by the German chemist Martin Heinrich Klaproth (1743–1817). Klaproth named the new element "titanium" after the Titans, who were a race of giant primordial deities in ancient Greek mythology.

The strong, low-density metal is ductile when pure and malleable when heated. Titanium compounds occur widely in nature. One sparkling example is how titanium dioxide (TiO_2) accounts for the asterisms observed in star rubies and sapphires. Industrial chemists incorporate titanium dioxide in paints, paper, and plastics. Engineers employ titanium and its alloys in aerospace systems because the metal has the highest strength-to-weight ratio of any known metal. It is as strong as steel but about 45 percent lighter, resists corrosion, and maintains its strength at elevated temperatures.

When heated to a temperature of 2,192°F (1,200°C [1,473 K]), pure titanium metal burns in air. It is the only element that will burn when exposed to a heated atmosphere of pure nitrogen gas. At 1,472°F (800°C [1,073 K]), titanium burns in nitrogen gas to form titanium nitride (TiN). Materials scientists use titanium nitride as a coating for cutting tools and drill bits.

Aluminum

Aluminum (Al) (British spelling aluminium) has an atomic number of 13 and an atomic weight of 26.982. This basic (or other) metal has a density of 168.6 lbm/ft^3 (2.70 g/cm^3) at room temperature and a melting point of 1,220.58°F (660.32°C [933.44 K]). At 68°F (20°C [293 K]), aluminum has a specific heat of 0.215 Btu/lbm-°F (0.90 kJ/kg-°C), a thermal conductivity of 136.9 Btu/(h-ft-°F) (237 J/[s-m-°C]), and an electrical conductivity of 37.67 $\times 10^6$ S/m.

Pure aluminum metal is silver-white in color, ductile, and relatively soft (Mohs hardness of 2.8), but the shiny metal's surface is quickly dulled by exposure to air, as a film of aluminum oxide (Al_2O_3) naturally forms on exposed surfaces. This thin layer of oxide actually helps the metal resist further corrosion. Because aluminum is ductile, the element can be easily drawn into wires or pressed into thin sheets or foil.

Although aluminum is the most abundant metal in Earth's crust, it is never found as a free metal in nature. All of Earth's aluminum has combined with other elements to form compounds. The rock bauxite consists of hydrated aluminum oxides and is the main ore of alumina (aluminum oxide), from which the metal is produced. In 1825, the Danish scientist Hans Christian Ørsted isolated tiny amounts of aluminum. Later that century in 1886, the American chemist Charles Martin Hall (1863–1914) and the French chemist Paul Louis Héroult (1863–1914) independently discovered a process for economically obtaining aluminum metal from alumina. Two years later, the Austrian chemist Karl Josef Bayer (1847–1904) developed a process to extract alumina from bauxite. Today, the Hall-Héroult and the Bayer processes still produce the bulk of the world's aluminum. Both processes are energy intensive and depend upon large supplies of inexpensive electricity.

Aluminum is the second most widely used metal in the world. Engineers use this strong, lightweight metal and its alloys in a wide variety of applications, including aircraft structures, rocket vehicles and spacecraft, automobiles, cans, kitchen utensils, and foils. Several new football stadiums in the National Football League (NFL) incorporate high-technology aluminum exteriors as part of their architectural design. Although aluminum does not conduct electricity as well as copper or silver, engineers use aluminum in long-distance electrical transmission lines because of its low mass. Aluminum metal is deposited on glass to make mirrors, and aluminum oxide is used to make synthetic rubies and sapphires for lasers.

CHAPTER 7

Building Materials

This chapter discusses some of the oldest and most durable solid materials that have been used by human beings. From the start of civilization, people have used various combinations of stone, clay, and sand to build permanent shelters, to pave roads, and to construct enduring monuments.

THE STONE AGE

The human use of solid materials began when a prehistoric ancestor of modern humans picked up a stone and started using it as a tool or weapon. The Paleolithic period (Old Stone Age) represents the longest phase in human development. Archaeologists and anthropologists often divide this large expanse of time into three smaller periods: the Lower Paleolithic period, the Middle Paleolithic period, and the Upper Paleolithic period. Taken as a whole, the most significant development that occurred during the Paleolithic period was the evolution of the human species from a near-human apelike creature into modern human beings *(Homo sapiens)*. The process was exceedingly slow, starting about 2 million years ago and ending about 10,000 years ago with the beginning of the Mesolithic period (Middle Stone Age), an era coinciding with the end of the last ice age.

During the Lower Paleolithic period (from approximately 2 million years ago until about 100,000 B.C.E.), early hunters and gatherers began to use simple stone tools for cutting and chopping. They also learned how to construct crude handheld stone axes. Caves and natural rock overhangs provided simple shelters. Irregular piles of stone may have formed protective barriers at the entrance of some caves. In the Middle Paleolithic period (from about 100,000 B.C.E. to about 40,000 B.C.E.), Neanderthals lived in caves, learned to control fire, and used improved stone tools for hunting, including spears with well-sharpened stone points. Neanderthals employed more carefully assembled stone walls and other barriers to protect and hide the entrances to their caves. They also used circles of stones to support wood frames and animal skin canopies for the temporary shelters that they occupied during hunting trips and seasonal migrations. For additional protection against the environment, these nomadic people learned how to use bone needles to sew furs and animal skins into body coverings.

It was during the Upper Paleolithic period (from about 40,000 B.C.E. to approximately 10,000 B.C.E.), that Cro-Magnons arrived on the scene and displaced Neanderthals. Cro-Magnon tribal clans engaged in better organized and more efficient hunting and fishing activities with the assistance of improved tools and weapons, including carefully sharpened obsidian and flint blades. A great variety of finely worked stone tools, better sewn clothing, the first human-constructed semipermanent shelters, and jewelry, such as bone and ivory necklaces, also appeared. Cro-Magnon peoples developed more elaborate fireplaces, adorned their cave walls with special paintings, and sometimes improved a cave's floor with a collection of inlaid rocks. The shamans of Cro-Magnon clans painted colorful animal pictures on the walls of caves. The ancient cave paintings that survive provide scientists some insight into the rituals of Cro-Magnon hunting cultures. One example is the interesting collection of cave paintings that were made sometime between 18,000 and 15,000 B.C.E. in Vallon-Pont-d'Arc, France.

As the glaciers of the last ice age thawed, the Mesolithic (Middle Stone Age) period featured the appearance of better cutting tools, carefully crafted stone points for spears and arrows, and the bow. Depending on the specific geological region, the Mesolithic period began some 12,000 years ago. This era of human development continued until it was replaced by another important technical and social transformation, the Neolithic (New Stone Age) Revolution. Scientists believe that humans domesticated the dog during the Mesolithic period and then began using the animal as a hunting companion. The prehistoric human-canine bond continues to this day.

The Neolithic Revolution was the incredibly important transition from hunting and gathering to agriculture. As the last ice age ended about 12,000 years ago, various prehistoric societies in the Middle East (the Nile Valley and Fertile Crescent) and in parts of Asia independently began to adopt crop cultivation. Since farmers tend to stay in one place, the early peoples involved in the Neolithic Revolution began to establish semi-permanent and then permanent settlements. During this period, people learned to use sand and clay to make pottery and bricks. The development of pottery supported the organized storage and transport of food and drink, while the development of clay bricks (sun-dried as well as fired) enabled the construction of individual habitats and the first cities.

Historians often identify this period as the beginning of human civilization. The Latin word *civis* means "citizen" or a "person who inhabits a city." The city of Jericho is an example. With a series of successive settlements ranging from the present back to about 9000 B.C.E., archaeologists regard Jericho, a town near the Jordan River in the West Bank of the Palestinian territories, as the oldest continuously inhabited settlement in the world. Stone cutting and grinding tools helped premetallurgical Neolithic people process agricultural harvests. Members of these early societies also employed stone to construct some of the world's more interesting ancient astronomical calendars (Stonehenge in Great Britain) and enduring monuments (the three large pyramids of Giza in Egypt).

Stonehenge is a collection of standing stones on the Salisbury Plain in the south central portion of Great Britain. Archaeologists currently believe that this prehistoric megalith was constructed in stages from 2750 B.C.E. to about 1550 B.C.E. In 1986, UNESCO named the site and its surroundings a world heritage site. A 300-foot (91 m)-diameter ditch encircles the prehistoric structure. Large stones are arranged in four distinct sets. The outermost set of stones consists of large sandstones about 13.5 feet (4.1 m) tall, which are connected by lintels. As used here, the term *lintel* refers to a long horizontal stone that spans an opening between two tall vertical rocks. Inside the outermost circle of large sandstones is a circle of blue stones, followed by a horseshoe-shaped arrangement of other stones, and then an innermost ovoid (egg-shaped) arrangement of stones. A large stone called the Altar Stone resides within the ovoid.

Scientists have proposed many theories about the purpose of Stonehenge, whose construction and active use appears to have extended from prehistoric times to the early Bronze Age in Great Britain. One prevalent theory is that Stonehenge was an ancient astronomical observatory/

calendar designed to monitor and predict the periodic motions of the Moon and Sun. More recent theories suggest the site may also have served as a center of healing or a cremation site for prehistoric tribal leaders and their families.

Modern materials science encounters ancient materials science. A U.S. Air Force B-1B Lancer bomber leads a flight of Egyptian, French, Greek, Italian, and U.S. aircraft over the Great Pyramids of Giza during Exercise Bright Star 1999–2000. *(U.S. Air Force)*

Bordering on the modern city of Cairo, Egypt, are the three large pyramids of Giza, the royal necropolis (burial city) for the pharaohs of the Fourth Dynasty of the Old Kingdom. The oldest and largest of these pyramids is called the Great Pyramid of Giza, or Pharaoh Khufu's Pyramid. Also called Cheops in ancient Greek, construction of Khufu's enormous pyramid was completed around 2560 B.C.E. For almost four millennia, the Great Pyramid remained the tallest human-made object on the surface of Earth. Erosion has now reduced the Great Pyramid's height to about 456 feet (139 m), but archaeologists suggest that the original height of this massive tomb was actually about 482 feet (147 m). Nearly equilateral triangles form the four exterior faces of Khufu's Pyramid, while its base forms an almost perfect square, 755 feet (230 m) on each side. Ancient Egyptian builders used mostly granite and limestone blocks to construct Khufu's Pyramid and the two smaller pyramids that dominate the Giza necropolis. Egyptian architects used more than 2 million giant stone blocks to construct Khufu's Pyramid. A typical limestone block used in this construction project had a mass of about two tons, that is, approximately 4,000 lbm (1,800 kg) each. Egyptologists estimate that some of the heaviest blocks used to create Khufu's burial chamber within the pyramid had an estimated mass of nine tons (18,000 lbm [8,165 kg]) each.

Pharaoh Khafre, who succeeded his father, Khufu, built his own (somewhat smaller) pyramid at Giza along with the Sphinx. The third and smallest of the three pyramids at Giza was constructed by Pharaoh Menkaure, thought to be the son of Khafre. As befitted a royal necropolis, the entire region around the three pyramids is dotted with secondary

away. Careful Roman engineering and the innovative use of cement and concrete made an aqueduct's flow channel water-tight. At the peak of the empire, a system of nine aqueducts carried millions of gallons (liters) of water to Rome each day. Urban life in Imperial Rome required an enormous quantity of water. Heated bathhouses served as a central social activity for most Roman citizens. These bathhouses often adjoined public toilet facilities. The latrines in the public toilets generally included marble seats and running water that continuously flushed human waste into a sanitary sewer system.

DIMENSION STONE AND CRUSHED STONE

Modern engineers and architects use two types of stone that are quarried, processed, and sold as commodities. These are dimension stone and crushed stone.

According to scientists at the U.S. Geologic Survey (USGS), *dimension stone* is defined as any type of natural rock material that is quarried in order to make slabs or blocks of rock that are cut to specific shapes and dimensions (that is, have a specific width, length, and thickness). As in ancient times, today's architects and engineers use dimension stone because it is strong, durable, and attractive. The most important rocks used as dimension stone are granite, marble, slate, limestone, and sandstone. For many decorative applications, dimension stone is polished.

More than half of the dimension stone quarried in the United States each year is rough block production. Architects and engineers use various rough block dimension stones in construction projects, including the foundations of monuments. Dressed dimension stones are more precisely cut and often highly polished, at least on facing surfaces. These stones often serve as flagstones for walkways or curbstones for streets. Architects select certain dimension stones to provide color and texture to a particular building or monument. The granite countertops found in many modern kitchens and bathrooms represent a common way people use dimension stone in their daily lives. Polished dimension stones are often used for tombstones, mausoleums, and monuments.

Art is perhaps the most notable application of dimension stone. Sculptors use dimension stone, such as a rough block of marble, to create an enduring work of art. The famous Renaissance sculptor Michelangelo Buonarotti took dimension stone quarried in Carrara, Italy, and chiseled the rough blocks of marble into incredibly beautiful works of art, such as the *Pietà,* completed in 1499, and the *Statue of David,* completed in 1504.

Geologists define *crushed stone* as any type of natural rock that has to first be blasted from its natural state in Earth's crust in order to be mined. The blasted rock is then crushed and screened to produce an appropriate size distribution. Limestone, dolomite, and granite are among the most common types of rocks processed into crushed stone. Marble, slate, quartzite, and sandstone also serve as crushed stone, but in much smaller quantities. The primary use for crushed stone is as aggregate for road construction and maintenance. Properly sized crushed stone forms a major component of the concrete used in road and building construction.

The quarrying of dimension stone *(USGS)*

The American Interstate Highway System—formally known as the Dwight D. Eisenhower National System of Interstate and Defense Highways—consists of almost 47,000 miles (75,640 km) of paved roads through December 2008. Natural aggregates (construction sand, gravel, and crushed stone) make up the largest component of nonfuel minerals consumed annually in the United States. To facilitate the construction of highways, engineers use natural aggregates to form a road base and also incorporate the aggregates into asphalt and concrete. Somewhat mimicking the practices of ancient Roman engineers, modern civil engineers construct a typical interstate highway as follows. They start with a bottom layer of well-compacted soil, add a 1.75-foot (0.53 m)-thick layer of natural aggregates, and then complete the highway by adding a 0.92-foot (0.28 m)-thick layer of concrete. By volume, pavement concrete is typically 60 to 75 percent aggregate, 15 to 20 percent water, 10 to 15 percent cement, and 5 to 8 percent entrained air.

The Federal Highway Administration within the U.S. Department of Transportation has established other standards for roads designated as part of the Interstate Highway System. Some of these standards include limited access, a road design that allows vehicle speeds of 50 to 70 mph (80 to 113 km/h), a minimum of two travel lanes in each direction, 12-foot (3.66 m) lane widths, a 10-foot (3.05 m) paved right shoulder, and a four-foot (1.22 m) paved left shoulder.

Much like the extensive road network of the ancient Roman Empire, the vast system of highways and roads now found in the United States is critical to the safety, economy, and happiness of the American people. The construction and maintenance of this modern network of roads and highways has involved and continues to require the use of large quantities of materials, including aggregates, asphalt, cement, and steel. In many parts of the United States, concrete from roads and structures as well as asphalt are being recycled and processed for use as crushed stone in new road construction projects.

MODERN CEMENT AND CONCRETE

Materials scientists define the word *cement* as any substance that can combine components or particles into an integrated solid or make previously separated surfaces adhere to one another. Within the world of building materials, the word *cement* is generally taken to mean portland cement (lowercase spelling is now preferred). The original name of this type of modern cement comes from a region of the United Kingdom in Portland, England. *Concrete* is the engineering world's most versatile and widely used human-made material and usually consists of a mixture of portland cement, coarse aggregate, fine aggregate, and water. Portland cement serves as the binding agent or ceramic glue that holds the ingredients of the concrete together.

A recipe for modern portland cement was patented in 1824 by the British brick mason and inventor Joseph Aspdin (1778–1855). His patent was entitled "An improvement in the modes of producing an artificial stone" and pertained to a product he called portland cement. This name refers to a rough-dimension stone quarried on the Isle of Portland along the British coast. Aspdin used the stone of Portland analogy to extol the qualities of his own cement mixture. The mixture and the name became part of the lexicon of modern engineering activities.

Today, the basic ingredients for portland cement are silica (SiO_2), alumina (Al_2O_3), lime (CaO), and iron oxide (Fe_2O_3). Materials scientists blend these materials together in appropriate proportions to manufacture different types of portland cement. Type I portland cement is the basic, general purpose cement that engineers typically use for sidewalks, reinforced concrete buildings, bridges, culverts, dams, and reservoirs. The varieties of portland cement extend up to Type V, which is a sulfate-resistant cement that engineers use when the concrete is exposed to high-sulfate-content soil or water. When water is added to portland cement, the mixture hardens and releases heat in a process call *hydration.* These hydration reactions are quite complex and still not completely understood by scientists. Hydration continues for months to years after the concrete and cement of a major structure are poured, further hardening and strengthening the structure as time passes.

Basically a composite ceramic material, concrete consists of portland cement, coarse aggregate, fine aggregate, and water. The cement paste (consisting of water and portland cement) forms a hard matrix that holds the aggregate particles together. The volume of a typical concrete consists of about 60 to 80 percent aggregates. The precise aggregate content and size distribution greatly impacts the final properties of the concrete. Engineers sometimes use air entrainment to make a particular concrete more workable or more resistant to freezing and thawing. For acceptable frost protection, *air-entrained concrete* generally contains between 4 and 8 percent air by volume. The entrained air is present in the form of very tiny bubbles typically with diameters on the order of 3.28×10^{-4} feet (100 μm) or less.

Concrete has a much higher compressive strength than tensile strength. Materials scientists design different concrete mixtures for different tasks and construction projects. Several factors that go into the choice of a concrete mixture include the strength and durability of the final product, the workability of the concrete during pouring and construction, and the cost. Engineers often cast concrete with embedded steel reinforcing bars or steel wire mesh. They call the resultant product *reinforced concrete.* It has increased tensile strength due to the embedded steel bars or wire mesh.

CHAPTER 8

Carbon—Earth's Most Versatile Element

Scientists define organic chemistry as the chemistry of compounds that contain carbon. They have discovered and identified more than 10 million carbon compounds, and the number continues to grow each day. This chapter describes why carbon is such a very special element. Carbon atoms have the rather unique ability to strongly bond with each other in a manner that supports the formation of very long chains. Many carbon compounds are essential to life on Earth. The chapter discusses the four principal allotropes of carbon: diamond, graphite, amorphous carbon, and fullerenes

CARBON—GENERAL CHARACTERISTICS AND APPLICATIONS

Carbon (C) has an atomic number of 6 and an atomic weight of 12.0107. This nonmetallic element has two stable isotopes, carbon-12 (natural abundance 98.89 percent) and carbon-13 (natural abundance 1.11 percent). With a half-life of 5,730 years, carbon-14 is the element's longest-lived radioisotope.

Carbon has two principal solid crystalline forms, diamond and graphite. Diamond is a very hard crystal that is typically colorless and prized for jewelry as well as being very useful in industrial applications that involve tough cutting operations. Graphite is a soft, black crystalline solid that is useful as a lubricant, for writing (the "lead" in pencils), and as electrodes in various electrical devices. Metallurgists and materials scientists have traditionally included coal, soot, charcoal, and coke in an allotropic category called amorphous carbon. This general categorization implies that amorphous carbon, though solid, has no distinctive large-scale crystalline structure. Finally, fullerenes represent the fourth major carbon allotrope—a nanotechnology world version of carbon that has incredibly interesting physical properties and offers very promising materials science applications this century. Diamonds, graphite, coal, and fullerenes are discussed in subsequent sections of the chapter.

Carbon was known by prehistoric peoples; the element's name comes from the Latin word *carbo,* meaning "charcoal." Metallurgists alloy iron with small quantities of carbon and other elements to make a very important and interesting variety of steel, the metal that empowers today's global civilization. Carbon compounds dominate the modern chemical industry, and hydrocarbons serve as the lifeblood of national and international transportation networks. Chemists define a *hydrocarbon* as any organic compound composed exclusively of carbon and hydrogen. Natural gas and petroleum are hydrocarbons that currently dominate world commerce and international politics. Polyethylene is a solid hydrocarbon used extensively in the manufacture of plastic products (see chapter 10). One of the most exciting frontiers in science is biochemistry—the study of how various compounds of carbon, hydrogen, nitrogen, oxygen, and other biogenic elements form the basis of all life on Earth. Finally, radioactive carbon-14 allows scientists to peek into the past.

THE MYSTERY OF DIAMONDS AND GRAPHITE

Why are a clear, sparkling diamond (the hard mineral that scratches glass) and a black-gray lump of graphite (the soft mineral that allows a pencil to leave a mark on paper) so different? This is a true mystery of nature because both objects consist of the very same substance, the chemical element carbon. Yet, despite their common elemental heritage, the diamond is one of the hardest known natural substances, while the mineral graphite is soft enough to be used as a lubricant or in a writing instrument. Furthermore, while diamond is a good electric insulator, graphite can conduct electricity. Scientists in the 18th and 19th centuries puzzled about these dramatic property differences of the same element without much success. It was only in the 20th century that they eventually obtained a

A magnificent collection of rough (top) and cut (bottom) diamonds *(© 2002 Gemological Institute of America [GIA]; used with permission)*

satisfactory explanation. The solution required an accurate understanding of what really lies at the heart of all matter—atoms.

As pure substances, the minerals diamond and graphite both contain only carbon atoms, but these carbon atoms are arranged in significantly different ways, thereby creating different molecular forms called allotropes. The carbon atoms in a diamond are held together by covalent bonds, which form a rigid three-dimensional crystal lattice that disperses light very well. Since every carbon atom in the diamond is tightly bonded to four other carbon atoms, the diamond has a rigid chain, or network, that results in one of the hardest known natural substances. Graphite is another allotrope of carbon in which covalent bonds form sheets of atoms in hexagonal patterns. However, the adjacent sheets of carbon atoms in graphite are more loosely bound to each other by much weaker intermolecular forces called van der Waals forces. The impact of this particular molecular arrangement is startling, since graphite is also a crystalline mineral, but one that is soft and brittle.

In ancient times, people found diamonds in the alluvial sands of India. Historians think that the famous Greek conqueror Alexander the Great (356–323 B.C.E.) encountered these beautiful natural gemstones as a result of military campaigns in India. Senior members of his army then introduced the gems to Macedonia and other parts of the ancient world shortly after his death. The Greeks recognized that diamonds were not only brilliant and dazzling but that they were also very hard. The English word *diamond* derives from the ancient Greek word *adamas* (αδαμας), meaning "invincible."

Diamond has a density of 219.1 lbm/ft³ (3.51 g/cm³), a specific heat capacity of 0.141 Btu/lbm-°F (0.59 kJ/kg-°C), and a thermal conductivity of 271.6 Btu/(h-ft-°F) (470 J/[s-m-°C]). Jewelers describe gem-quality diamonds using four characteristics: carat, cut, color, and clarity. The mass of a gem-quality diamond is traditionally expressed in carats, with each carat mass further divided into 100 points. The cut of a diamond refers to how successfully an artisan cuts the stone's facets to achieve maximum brilliance. When properly cut and polished, the diamond's refractive index and ability to disperse light gives the crystalline gem its fire, or brilliance. Some of the more popular types of cut and polished diamonds are round diamonds, pear-shaped diamonds, radiant-cut diamonds, and oval diamonds. Diamond is generally colorless, although the presence of impurities can result in gems that exhibit a variety of subtle colors. Natural yellow, pink, blue, and green diamonds have been found. Lapidaries cut and polish these colored diamonds into very unusual and attractive gems. *Clarity* is a term gemologists use to describe a diamond's purity. Natural diamonds formed deep within Earth's crust usually have small traces of other elements trapped within. If too many tiny impurities are trapped within the diamond, then the gem has poor clarity.

With a value of 10 on the Mohs hardness scale for minerals, scientists have traditionally regarded diamond as the hardest known natural substance. Until materials science developments within the last decade, that perception was quite correct. Today, several other materials have seized the hardness title from the historic champion. One such material is called *lonsdaleite* (or hexagonal diamond). This substance, discovered in nature in 1967, is another allotrope of carbon that forms from graphite in meteorites when they impact Earth's surface at great velocity. Materials scientists speculate that pure samples of lonsdaleite should exhibit a level of hardness about 50 percent greater than that of diamond. Another contender is a compound known as wurtzite boron nitride (w-BN). This material is structured like diamond but does not contain carbon atoms. Wurtzite boron nitride is formed under the high pressure and temperature conditions of a volcanic eruption. Scientists speculate that the extra hardness of both lonsdaleite and wurtzite boron nitride is due to the fact that both substances exhibit some flexibility of their chemical bonds when experiencing stress. Since both lonsdaleite and wurtzite boron nitride are

WORLD'S LARGEST DIAMOND

In 1898, the South African diamond entrepreneur Sir Thomas Cullinan (1862–1936) discovered the Premier diamond fields. He eventually purchased the land and began exploiting the Premier mine. The largest diamond yet discovered is the Cullinan diamond, found in the Premier mine in South Africa in 1905. The rough diamond had an estimated mass of 3,100 carats. The South African government purchased this enormous rough diamond in 1907 and presented it to the reigning British monarch, King Edward VII. The stone was then carefully cut into nine large stones and about 100 smaller ones. The largest of these is the magnificent 530-carat, pear-shaped diamond called the Great Star of Africa (or Cullinan I). This flawless and colorless diamond is set in the British monarch's royal scepter. The modern Imperial British Crown of State contains a total of 2,783 diamonds, 17 sapphires, 277 pearls, 11 emeralds, and 5 rubies, including the very large 317-carat Cullinan II diamond (or the Second Star of Africa).

quite rare in nature, diamond remains popularly recognized as the hardest natural substance.

In the 1950s, scientists learned how to synthesize diamonds by subjecting graphite to high pressures and temperatures under controlled laboratory conditions. Only a few gem-quality diamonds were actually synthesized, so they could not economically compete with mined natural diamonds. The real commercial application of synthetic diamonds lies in the diamond films found on industrial cutting and polishing tools. Inferior mined natural diamonds (those not of gem quality) as well as human-made miniature diamonds have great industrial value in cutting, drilling, grinding, and polishing tools. Scientists involved in diamond synthesis research also use chemical vacuum deposition processes to place diamond films on various surfaces, including microelectronic components.

In sharp contrast to diamond's hardness, graphite is one of the softest materials known (Mohs mineral hardness of about 0.5–1.0). Graphite has a density of 141.5 lbm/ft^3 (2.267 g/cm^3) at room temperature, a specific heat of 0.170 Btu/lbm-°F (0.71 kJ/kg-°C), and a thermal conductivity ranging from 14.4 to 271.6 Btu/(h-ft-°F) (25 to 470 J/[s-m-°C]). At normal atmospheric pressures, graphite sublimes at about 6,422°F (3,550°C [3,823 K]). A person's most common encounter with graphite is usually as the so-called lead in a common pencil. Graphite has numerous industrial and scientific uses, such as a dry lubricant, in batteries, as the electrodes in electric arc furnaces and metallurgical processing operations, as the refractory surface in high-temperature crucibles and furnaces, in moderator rods in nuclear power reactors, and in specialized aerospace applications. Engineers use graphite to manufacture long-lifetime, high-performance carbon brushes in electric motors.

COAL

Coal, the rock that burns, represents one of the world's major energy resources. Coal is the fuel that provides the thermal energy necessary to generate more than half of the electricity used in the United States.

As shown in the accompanying illustration, coal forms from decayed and compacted plant remains. The first stage of coal formation is the creation of peat, which is partially decayed plant material that is protected from further oxidation (decay) by submersion in water. Over time, as peat is subjected to increasing lithostatic pressure and temperature, it gradually transforms under these geologic forces into coal in the following sequence: lignite, subbituminous coal, bituminous coal, and finally anthracite.

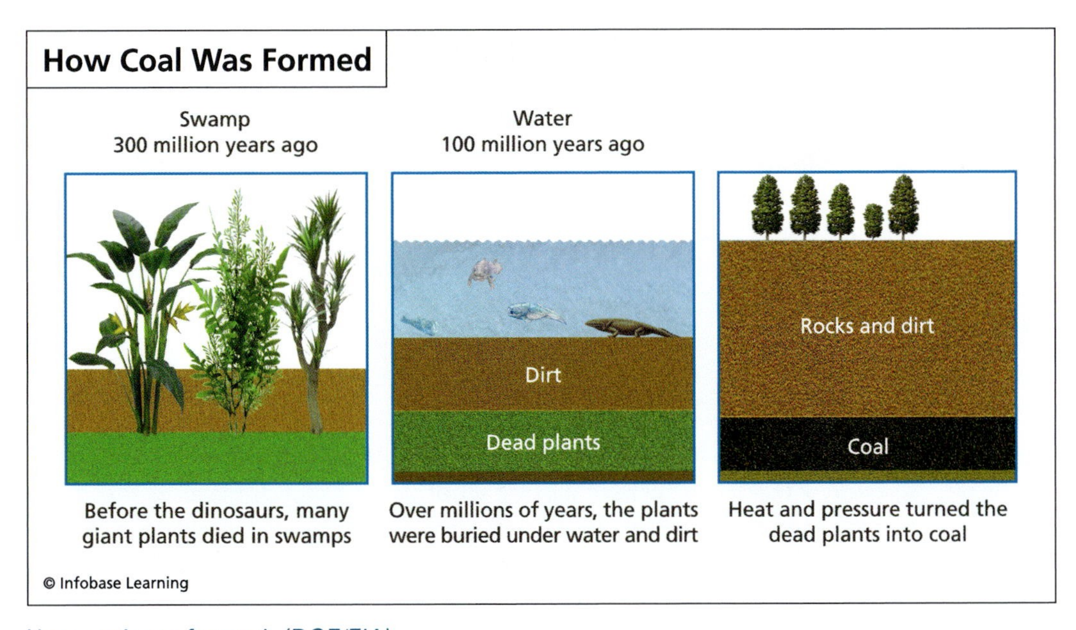

How coal was formed *(DOE/EIA)*

Lignite forms first, as increased pressures and heat within Earth's crust cause buried peat to dry and harden. It is a brown-colored coal with high moisture and ash content and relatively low heating value (about 4,000–8,300 Btu/lbm [9.3–19.3 MJ/kg]). Lignite has a relatively low carbon content, which ranges from about 25 to 35 percent by weight.

As the geologic pressure increases, lignite transforms into the next type of coal. Subbituminous coal is dull black in color and has a carbon content ranging from 35 to 45 percent. This type of coal has a heating value ranging from 8,300 to 13,000 Btu/lbm (19.3–30.2 MJ/kg). Although subbituminous coal has a lower heating value than bituminous coal or anthracite, utilities often prefer to burn it in their electric power generating stations because of its generally lower sulfur content.

At even greater geologic pressure, subbituminous coal transforms into bituminous coal. Also known as *soft coal,* bituminous coal is the most plentiful type of coal found in the United States. This type of coal is used primarily to generate electricity and to manufacture coke for the steel industry. *Coke* is the porous but hard gray-colored carbon fuel that results when bituminous coal is heated to high temperatures under tightly controlled air flow conditions intended to drive off easily vaporized contaminants. Bituminous coal has a carbon content that ranges from 45 to 86 percent and a heating value that ranges from about 10,500 to 15,000 Btu/lb (24.4–34.9 MJ/kg).

Sometimes called *hard coal,* anthracite forms when tectonic processes place bituminous coal under conditions of great geological pressure, such as when mountain ranges form. Anthracite coal has the highest carbon content of all coals (up to 98 percent) and the highest heat value (about 15,000 Btu/lb [34.9 MJ/kg]). In the United States, anthracite coal deposits are found only in the northeastern portion of Pennsylvania. This form of coal is used for the generation of electric power and also for heating homes.

CARBON CYCLE

Carbon is the basis for life on Earth. This essential element moves through the planet's biosphere in a great act of natural recycling called the *global carbon cycle.* Scientists find it helpful to divide the global carbon cycle into two components: the *geologic carbon cycle,* which operates over very long time periods (millions of years); and the *physical/biological carbon cycle,* which takes place over much shorter periods (days to millennia).

The carbon that now cycles through Earth's various planetary systems (that is, the biosphere, atmosphere, hydrosphere, and lithosphere) was present in the ancient solar nebula from which the solar system began to form about 5 billion years ago. As Earth emerged as a distinct celestial object about 4.6 billion years ago, the young planet's surface was extensively bombarded by carbon-bearing meteorites. Over time, these meteorite impacts steadily increased the planet's carbon content. Since the end of the period of great cosmic collisions, carbonic acid has slowly and steadily combined with calcium and magnesium in Earth's crust to form carbon-containing chemical compounds called carbonates. Carbonic acid (H_2CO_3) is a weak acid that forms when gaseous carbon dioxide (CO_2) dissolves in water (H_2O). Through weathering, the carbonic acid combined with the calcium and magnesium found in Earth's crust to form chemical compounds such as calcium carbonate (limestone). This process was continual but very gradual. Another natural process called *erosion* washed the carbonates into the ancient oceans. Once in the ocean, these carbon-bearing compounds precipitated out of the ocean water and formed layers of sediment on the ocean floor.

The process of plate tectonics then pushed these carbon-bearing sediments beneath the continents. Geologists refer to this activity as *subduction.* Once deep within the lithosphere, the limestone and other carbon-bearing sediments experienced increased heat and pressure. The carbonates melted, reacted with other minerals, and released carbon dioxide. Volcanic eruptions then returned the released carbon dioxide to Earth's atmosphere. Scientific evidence suggests that there is a natural balance in the geologic carbon cycle between weathering, erosion, subduction, and volcanism. The geologic carbon cycle still regulates atmospheric carbon dioxide concentrations, but it operates over time periods of hundreds of millions of years.

In the physical/biological carbon cycle, living things play a major role in moving carbon through the biosphere. During the process of photosynthesis, green plants absorb carbon dioxide from Earth's atmosphere and use sunlight (energy) to create fuel (glucose and other sugars) for constructing more complex plant structures. Scientists describe *photosynthesis* as the natural process through which chlorophyll-bearing green plants use the energy content of sunlight to make carbohydrates from atmospheric carbon dioxide and water. In the basic process, plants transform input hydrogen, oxygen, and carbon into the stable organic compound glucose ($C_6H_{12}O_6$) and release water and oxygen back into Earth's environment.

Chemists define *carbohydrates* as any member of a large class of carbon-bearing compounds, including sugars, starches, cellulose, and similar compounds. Carbohydrates serve as the enabling fuel for living things and allow organisms to grow and reproduce. Plants and animals metabolize (burn) carbohydrates and other nutrient molecules. Through aerobic *respiration,* the opposite of photosynthesis, living organisms release the energy stored in carbohydrates by combining nutrient organic molecules such as glucose with oxygen and produce water and carbon dioxide. The released carbon dioxide carries carbon back into the atmosphere. Another natural process, *decomposition* (the digestion of dead or decaying organic matter by bacteria and fungi), also returns carbon fixed by photosynthesis back into the atmosphere. Carbon circulates through the biosphere because of the linkage between photosynthesis, respiration, and decomposition.

Photosynthesis and respiration play important roles in moving carbon through the biosphere on time scales far shorter than those involved in the geologic carbon cycle. The biological processes for various living organisms are elegantly complex; only the very basic carbon exchange activities are mentioned here. Scientists currently estimate that the yearly quantity of carbon fixed by photosynthesis and released back to the atmosphere by respiration is about 1,000 times greater than the amount of carbon that moves through the geologic carbon cycle on an annual basis.

Millions of years ago, some of the carbon involved in biological processes was not released back into the atmosphere as carbon dioxide. Instead, buried deposits of dead plants on land and certain marine life forms in the oceans became compressed over time by layers of sediment, eventually forming fossil fuels such as coal, oil, and natural gas. The carbon locked in these fossil fuels remained trapped within Earth's crust for millions of years until humans mined the various fuels and began burning them. Since the start of the Industrial Revolution in the 17th century, the carbon dioxide content of Earth's atmosphere has increased from about 280 parts per million (ppm) to a current (2010) value of about 385 ppm. Human activities, especially the consumption of enormous quantities of fossil fuels and large-scale deforestation, now exert a measurable influence on the planet's global carbon cycle. Many scientists are alarmed at the increased quantity of carbon dioxide in the atmosphere and warn about the dire consequences of global warming due to greenhouse gas buildup. Climate models indicate that increased

CARBON SEQUESTRATION

Energy demand projections by the U.S. Department of Energy in 2010 indicate that fossil fuels will continue to serve as the major source of energy throughout the world for many decades. Since the consumption of fossil fuels remains intimately linked to national security and economic vitality, scientists are very busy investigating ways to keep the atmospheric concentration of carbon dioxide from rising. One approach to controlling carbon emissions from fossil fuels is called carbon sequestration.

Scientists define *carbon sequestration* as the capture and long-term storage of carbon dioxide and other greenhouse gases such as methane that would otherwise enter Earth's atmosphere. They suggest the greenhouse gases can either be captured at the point of origin (direct sequestration) or else removed from the atmosphere (indirect sequestration). The captured carbon dioxide can then be stored in underground reservoirs (geologic sequestration), injected into deep portions of the oceans (ocean sequestration), or stored in vegetation and soils (terrestrial sequestration).

(continues)

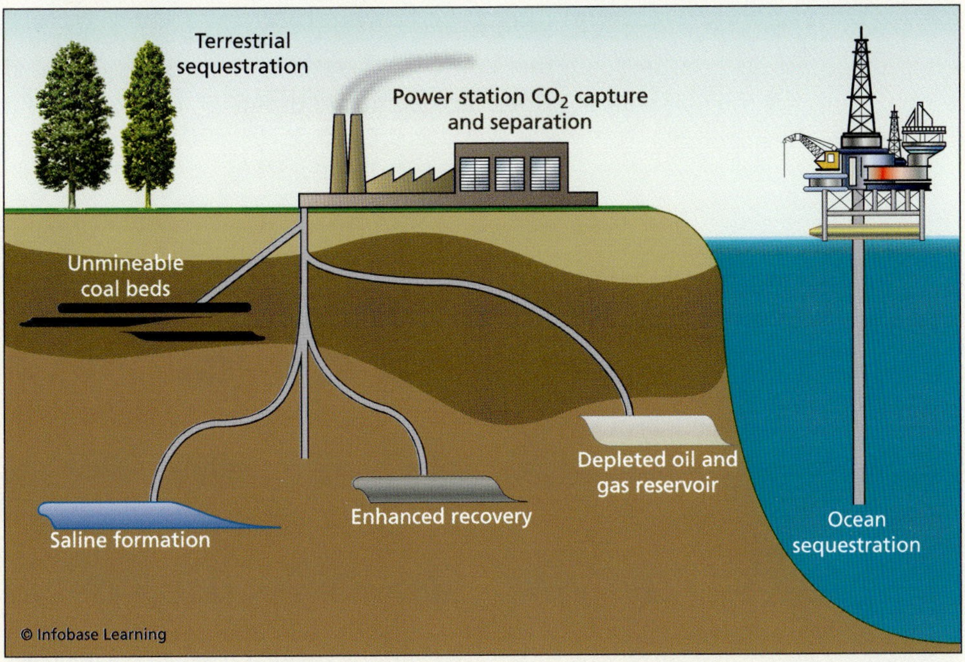

This illustration depicts several possible approaches to carbon sequestration. ***(DOE/EIA)***

(continued)

Geologic sequestration involves the storage of captured carbon dioxide in depleted oil and gas reservoirs, in coal seams that can no longer be practically mined, or possibly within underground saline formations. The high-pressure injection of carbon dioxide into depleted oil and gas reservoirs may also force any remaining oil or gas toward production wells, facilitating enhanced hydrocarbon recovery operations. Ocean sequestration involves directly injecting carbon dioxide deep into the ocean. Although carbon dioxide is soluble in sea water and the oceans naturally absorb and release huge amounts of carbon dioxide, this proposed technique is not without controversy. The controversy centers around what impact deep-water injection activities might have on the ocean and various marine ecosystems. Terrestrial sequestration involves removal of carbon dioxide from the atmosphere by means of vegetation and soils. Ecosystems that offer significant opportunities for enhanced carbon sequestration include forests, biomass crops, grasslands, and peat lands.

Advanced sequestration concepts, such as converting captured greenhouse gases into rocklike solid materials, are also being investigated. Scientists are examining the feasibility of using minerals such as magnesium carbonate for carbon capture and storage. Before any carbon sequestration approach is implemented on a large scale this century, scientists and engineers must demonstrate that the selected approach is technically practical, economic, and environmentally acceptable.

greenhouse gas concentrations (including methane [CH_4] and carbon dioxide) have been the primary driver of Earth's increasing surface temperature. Improved ways of tracing the amount and pathways of carbon as this element travels throughout Earth's biosphere is an important aspect of Earth systems science. Such scientific activities should provide data that lead to a more responsible level of stewardship of humans' home planet in this century.

FULLERENES

A significant milestone in modern carbon chemistry occurred in 1985, when researchers Richard E. Smalley (1943–2005), Robert F. Curl, Jr.

(1933–), and Sir Harold W. Kroto (1939–) discovered an amazing molecule that consisted of 60 linked carbon atoms (C_{60}). Smalley named these distinctive clusters of carbon atoms *buckyballs* in honor of the famous architect Buckminster Fuller (1895–1983), who promoted the use of geodesic domes. Scientists had known for centuries that carbon consisted of two allotropes, namely diamond and graphite, so the collaborative discovery of another carbon allotrope quite literally rocked the world of chemistry.

Buckyballs are very hard to break apart. When slammed against a solid object or squeezed, they bounce back. The cluster of 60 carbon atoms is especially stable. It has a hollow, icosahedral structure in which the bonds between the carbon atoms resemble the patterns on a soccer ball. Smalley, Curl, and Kroto shared the 1996 Nobel Prize in chemistry for their "discovery of carbon atoms found in the form of a ball."

Identification of this new allotrope of carbon sparked broad interest in the chemistry of an entire family of hollow carbon structures, now referred to collectively as *fullerenes.* Formed when vaporized carbon condenses in an inert atmosphere, fullerenes consist of a wide range of shapes

This computer-generated illustration shows that buckyballs, discovered in space by NASA's *Spitzer Space Telescope,* closely resemble black-and-white soccer balls, but on a much smaller scale. *(NASA/JPL-Caltech)*

and sizes, including *nanotubes,* which are of interest in electronics and hydrogen storage. Nanotubes are cylindrically shaped tubes of carbon atoms about 3.28×10^{-9} feet (1 nm) in diameter. They are stronger than steel and can conduct electricity. The walls of nanotubes have the same soccer ball–like structure as buckyballs, but they come rolled up into long tubes. Fullerenes represent a primary research area in contemporary nanotechnology. Since they come in many variations, highly versatile fullerenes promise many potential applications. For example, researchers believe that fullerene structures can be manipulated to produce superconducting salts, new catalysts, new three-dimensional polymers, and biologically active compounds.

In the 21st century, as humans learn to manipulate matter at the atomic level (nanotechnology), new, more exciting applications of solid matter will influence the trajectory of civilization. Nanotechnology represents an extremely important plateau in the history of civilization—somewhat analogous in its overall potential impact to the discovery and use of fire by humans' prehistoric ancestors. What is clearly different in this case is the timescale within which the benefits and risks associated with future discoveries can impact humans' global civilization. Sweeping technical changes in matter manipulation and materials science could occur in periods as short as decades or possibly even a few years. Such radical changes in the manipulation of matter will undoubtedly be accompanied by inevitable stresses on existing environmental, social, and economic infrastructures.

There are enormous technical challenges facing the scientists and engineers who currently labor at the frontiers of nanotechnology. Many social, ethical, political, and economic issues also need to be carefully examined before laboratory discoveries are released to widespread application. Unbiased and careful assessments of potential risks are necessary before human-engineered nanoscale devices are released in large quantities into the general environment or the human body. Particular attention must be given to any molecular manufacturing process that involves self-replicating nanoscale systems. Protocols and safeguards must be well established before the application of such devices or systems transitions from laboratory demonstration to large-scale industrial or commercial applications. The price of harvesting huge benefits from this anticipated era of rapid technical progress in materials science is heightened social and ethical vigilance.

RADIOCARBON DATING

The American chemist Willard Frank Libby (1908–80) employed the phenomenon of radioactivity to unlock the secret of time by developing the method of carbon-14 dating. His discovery provided an important research tool to archaeology, geology, and other branches of science.

Libby was born on December 17, 1908, in Grand Valley, Colorado, and graduated from the University of California at Berkeley with a Ph.D. in chemistry in 1933. While performing his doctoral research, Libby constructed one of the first Geiger-Muller tubes in the United States. The device could detect certain forms of nuclear radiation, and the experience prepared Libby for his Nobel Prize–winning research in the late 1940s.

After World War II in 1945, Libby accepted a position as professor of chemistry at the Institute for Nuclear Studies at the University of Chicago (now the Enrico Fermi Institute for Nuclear Studies). It was there in 1947 that he developed the innovative concept for the carbon-14 dating technique, a powerful research tool for reliably dating objects up to about 70,000 years old (with today's equipment). He based the clever technique on the decay of the radioisotope carbon-14, as contained in such formerly living organic matter as wood, charcoal, parchment, shells, and even skeletal remains, and on the assumption that the absorption of atmospheric carbon-14 (produced by cosmic ray interactions in Earth's atmosphere) ceases when a living thing dies. At that point, the radioactive decay clock in the organic matter starts ticking. Scientists can determine an object's age by comparing its reduced carbon-14 activity levels to those higher carbon-14 activity levels found in comparable living organisms or viable organic materials.

In developing his idea, Libby reasoned that the level of carbon-14 radioactivity in any piece of organic material should clearly indicate the time of the organism's death. The real challenge he faced and overcame was to develop and operate a sensitive enough radiation detection instrument that was capable of accurately counting the relatively weak beta decay events of carbon-14. With a half-life of 5,730 years, radiocarbon is only mildly radioactive.

Libby constructed a sufficiently sensitive Geiger counter. He then successfully tested his proposed carbon dating technique against organic objects from antiquity, objects that had reasonably well-known ages. For example, he successfully used radiocarbon dating to determine the ages of a wooden boat from the tomb of an Egyptian pharaoh, a piece of prehistoric sloth dung that was found in Chile, and a wrapping from the Dead Sea Scrolls. His research demonstrated that carbon-14 analysis represented a reliable way of dating organic objects from the past. Libby received the 1960 Nobel Prize in chemistry for developing the method of carbon-14 dating and for its many important applications in archaeology, geology, and other branches of science. He summarized this interesting use of carbon in the 1952 book *Radiocarbon Dating.*

CHAPTER 9

Sand, Silicon, and Ceramics

This chapter describes how common solid materials such as sand and clay supported the advancement of civilizations throughout human history. The importance of these materials continues to the present day. In the second half of the 20th century, physicists and engineers began using high-purity silicon to make the semiconductors that make possible advanced electronic devices and computers. Their creative manipulation of silicon enabled the digital information age. An industrial area in California's Santa Clara Valley, about 20 miles (32 km) long and situated between the cities of Palo Alto and San Jose, is now known throughout the world as Silicon Valley. The term *Silicon Valley* has also become synonymous with leading-edge, high-technology endeavors.

SAND

Some of the happiest childhood memories for many people occurred building sand castles at the beach. The primary component of typical beach sand is quartz, also called silicon dioxide (SiO_2) or *silica.* Silicon dioxide is the most abundant chemical compound in Earth's crust.

Geologists describe the breakdown of surface rocks into smaller and smaller fragments using the following descriptive order from larger to smaller fragments: boulders, cobbles, gravel, sand, silt, and microscopic clay-sized particles. The finer the particle, the easier it is for natural forces

One of the joys of childhood is building a sand castle during a visit to the beach. Typical beach sand is quartz, or silica (silicon dioxide [SiO_2]). *(EPA)*

to move it. Within the field of geology, a particle (grain) of sand has a diameter that ranges from 0.00246 in (0.0625 mm) to 0.0787 in (2.0 mm). Therefore, sand particles are smaller than gravel but larger than particles of silt and clay. (Clay is discussed later in the chapter.) All rocks (igneous, metamorphic, and sedimentary) are subject to weathering and decomposition. Sediments are transported by water or wind erosion, and rock fragments experience the dissolution and transport of their soluble chemical components by surface water and groundwater.

Quartz is a hard mineral that does not fracture easily. The quartz minerals contained in granitic rocks typically range in color from clear to white, but they can also display any color from the iridescence of opal to the black of flint. Some sedimentary rocks, such as sandstone, have a large amount of quartz incorporated within them. Depending on how far the sand particles must travel before they reach the ocean, they may possess somewhat sharpened angular edges, or else they may have become very rounded and smooth. The smoothness of sand particles determines the overall physical properties of this granular material, including porosity and permeability.

Scientists define *porosity* as the ratio of pore (void) volume to total volume of a granular substance. The more porous a substance is, the more voids, or empty spaces, per unit volume it has. Beach sand and sandy soils have large voids and, therefore, higher values of porosity than clays and other fine-grained soils. Scientists define *permeability* as the ability of a porous solid to allow the passage of a fluid through it when the flowing fluid experiences a (hydraulic) pressure differential. An impermeable material has poorly connected voids and so greatly restricts or totally prohibits the passage of a fluid. When the voids of a porous rock or granular material are well interconnected, a fluid can flow through the substance with relative ease. In sandstone, the well-rounded sand grains create an ample number of interconnected void spaces, so sandstone represents a high-porosity and a high-permeability solid material.

From antiquity, people have used sand in a number of important ways. Some of these applications include the production of mortar, cement, concrete, bricks, plasters, and paving materials. Artisans have used sand in the manufacture of pottery and glass. Engineers employed sand to filter impurities out of water and other fluids. They also used sand as an inexpensive abrasive material. The term *sandblasting* describes the technique of propelling grains of sand at high velocity to clean or etch a surface.

Sand and silicon continue to play a major role in sustaining today's information-based global civilization. Modern chemical engineers produce commercial quantities of the element silicon (Si) by reacting sand (silica) and carbon in a high-temperature electric furnace. Ingots of ultrapure silicon support the manufacture of semiconductors. Of all the many exciting new materials that appeared in the 20th century, none has transformed the daily lives of so many people to the same extent as silicon-based semiconductors.

SILICON

Silicon (Si) is the seventh most abundant element in the universe and the second most abundant element found in Earth's crust. The element oxygen is the most abundant element in Earth's crust (46.1 percent by weight), followed by silicon (28.2 percent by weight). These two elements combine to form the compound silicon dioxide (SiO_2), the most abundant chemical compound contained in Earth's crust (42.86 percent by weight).

Silicon was first isolated in 1824 by the Swedish chemist Jöns Jacob Berzelius (1779–1848). The metalloid element's name derives from *silicis,* the Latin word meaning "flint." Silicon has an atomic number of 14 and an atomic weight of 28.0855. At 68°F (20°C [293 K]), silicon has a density of 145.4 lbm/ft^3 (2.33g/cm^3). At room temperature, the element exists in two allotropes, amorphous and crystalline.

Amorphous silicon is a brown powdery substance, while crystalline silicon has a grayish color with a metallic luster. Industrial chemists produce commercial quantities of silicon by heating sand (SiO_2) with carbon in special electric furnaces where temperatures reach 3,992°F (2,000°C [2,473 K]). At one atmosphere pressure, silicon has a melting point of 2,577°F (1,414°C [1,687 K]). Silicon has the following physical properties at room temperature: a specific heat of 0.170 Btu/lbm-°F (0.71 kJ/kg-°C) and a thermal conductivity of 86.1 Btu/(h-ft-°F) (149 J/[s-m-°C]).

Materials scientists have learned how to grow ultrapure single crystals of silicon. When doped with minor amounts of selected impurities, such as boron, gallium, germanium, phosphorous, or arsenic—high-purity silicon becomes very useful in the manufacture of solid-state electronic devices.

Scientists recognize that some substances, such as silver and copper, readily conduct the flow of electricity and call these materials *electrical conductors.* Other materials, such as wood and most plastics, resist the flow of electrons and are called *electrical insulators.* The basic difference between conductors and insulators lies in their atomic structures. A material that is a good electrical conductor has one or several outer (valence) electrons that are loosely attached to the parent atomic nucleus and, therefore, available to freely wander through the material under the influence of an applied voltage. The situation is quite different for a material that is a good electrical insulator. In this case, almost every electron remains tightly bound to its parent atom, so even when a voltage difference is applied across the material, the outer electrons cannot wander freely through it.

When scientists discuss the property of electrical conductivity, they also include a third common type of material called a *semiconductor.* A semiconductor is a solid crystalline material, such as silicon or germanium, that has a typical electrical conductivity intermediate between the values of good electrical conductors and insulators. Semiconductors can carry electric charges, but do not do so very well. The conductance of silicon is about 1 million times less than the electrical conductance of copper.

The electrical resistivity (symbol ρ) of a material is the reciprocal of that material's electrical conductivity. Silver has an electrical resistivity of 1.6×10^{-8} ohm-meter (Ω-m) at room temperature. By comparison, at 68°F (20°C [293 K]), pure silicon has an electrical resistivity value of 2,500 Ω-m; a typical n-type silicon semiconductor material has an electrical resistivity of 8.7×10^{-4} Ω-m; and a typical p-type silicon semiconductor material has an electrical resistivity of 2.8×10^{-3} Ω-m. Scientists call silicon that has been doped with (electron) donor atoms, such as phosphorus, *n-type semiconductors,* and silicon that has been doped with (electron) acceptor atoms, such as aluminum, *p-type semiconductors.* Finally, scientists and engineers recognize that glass and fused quartz are good electrical insulators. The electrical resistivity of glass at 68°F (20°C [293 K]) ranges from 10^{10} to 10^{14} Ω-m. Fused quartz has an approximate electrical resistivity of 10^{16} Ω-m.

The digital information age is a generic expression that actually encompasses several major information technology (IT) shifts that occurred in the mid- to late 20th century. The first part of this process was the discovery of the transistor in the late 1940s. The second part involved the application of the transistor in the subsequent microelectronics revolution. The third part of the IT revolution involved the development of the integrated circuit (IC). The modern integrated circuit represents nothing short of a technical miracle in materials science and engineering. The availability of inexpensive, reliable IC devices accelerated the exponential development and application of digital computers and microprocessors. Information technology specialists define a *microprocessor* as an integrated circuit or collection of integrated circuits contained on a single semiconductor chip and capable of performing most, if not all, of the functions of a digital computer's central processing unit (CPU). The CPU is the computational and control unit of a digital computer—the device that functions as the computer's "brain." The CPU interprets and executes instructions and transfers information within the computer. Microprocessors contain single-chip CPUs, while the CPUs in large mainframe computers contain numerous circuit boards, each packed full of integrated circuits.

In addition to the very important applications just discussed, silicon forms many other chemical compounds that are useful in today's world. With a value of approximately 7 on the Mohs hardness scale, crystalline silicon is a relatively hard but brittle substance. However, the compound silicon carbide (SiC) is nearly as hard as diamond and is used extensively as an abrasive. Industrial chemists employ sodium silicate (Na_2SiO_3) in the production of soaps and adhesives. Military engineers use silicon tetrachloride ($SiCl_4$) to create smoke screens.

Silicon is also a key ingredient in a large class of polymeric compounds called *silicones*. Chemists have created many different silicone materials that support a wide range of applications, including as lubricants, electrical insulators, medical implants, polishing agents, nonstick cookware, sealants, and adhesives. (Polymers are discussed in chapter 10.) Finally, silica (sand) is the principal ingredient of glass, a generally transparent or sometimes translucent substance that scientists consider an amorphous solid. Archaeologists suggest that the first human-made glass was manufactured in ancient Egypt about 5,000 years ago. Artifacts indicate that the ancient Egyptians mixed sand (silica [SiO_2]), calcium carbonate (limestone [$CaCO_3$]), and sodium carbonate (soda ash [Na_2CO_3]) to make glass. They heated the solid ingredients until they melted; stirred the molten

INTEGRATED CIRCUIT (IC)

An integrated circuit (IC) is an electronic circuit that includes transistors, resistors, capacitors, and their interconnections—all fabricated on a very small piece of semiconductor material (usually referred to as a *chip*). The American electrical engineer Jack S. Kirby (1923–2005) invented the integrated circuit while he was working on electronic component miniaturization during the summer of 1958 at Texas Instruments. Although Kirby's initial device was quite crude by today's standards, it proved to be a groundbreaking innovation that paved the way for the truly miniaturized electronics packages that now define the digital information age. Kirby shared the 2000 Nobel Prize in physics "for his part in the invention of the integrated circuit."

As sometimes happens in science and engineering, two individuals independently came up with the same innovative idea at about the same time. In January 1959, another American engineer, Robert N. Noyce (1927–90), while working for a California company named Fairchild Semiconductor, independently duplicated Kirby's feat. In 1971, Noyce, who was then the president and chief executive officer of another Silicon Valley company called Intel, developed the world's first microprocessor. Today, both Kirby and Noyce are recognized as having independently invented the integrated circuit.

The category of an integrated circuit, such as LSI or VLSI, refers to the level of integration and denotes the number of transistors on a chip. Using one common (yet arbitrary) standard, engineers say a chip has small-scale integration (SSI) if it contains less than 10 transistors, medium-scale integration if it contains between 10 and 100 transistors, large-scale integration (LSI) if it contains between 100 and 1,000 transistors, and very large-scale integration (VLSI) if it contains more than 1,000 transistors.

mixture carefully until it became a viscous, homogeneous liquid; and then carefully poured the molten mixture into appropriately shaped molds or forms. Rapid cooling allowed the molten mixture to become a transparent solid. Scientists call this type of glass *soda lime glass.*

By adjusting the proportions of the basic ingredients and adding or substituting small quantities of other materials, such as lead, boron, chromium, or cobalt, materials scientists learned how to manufacture many different types of glasses, each with special properties and applications.

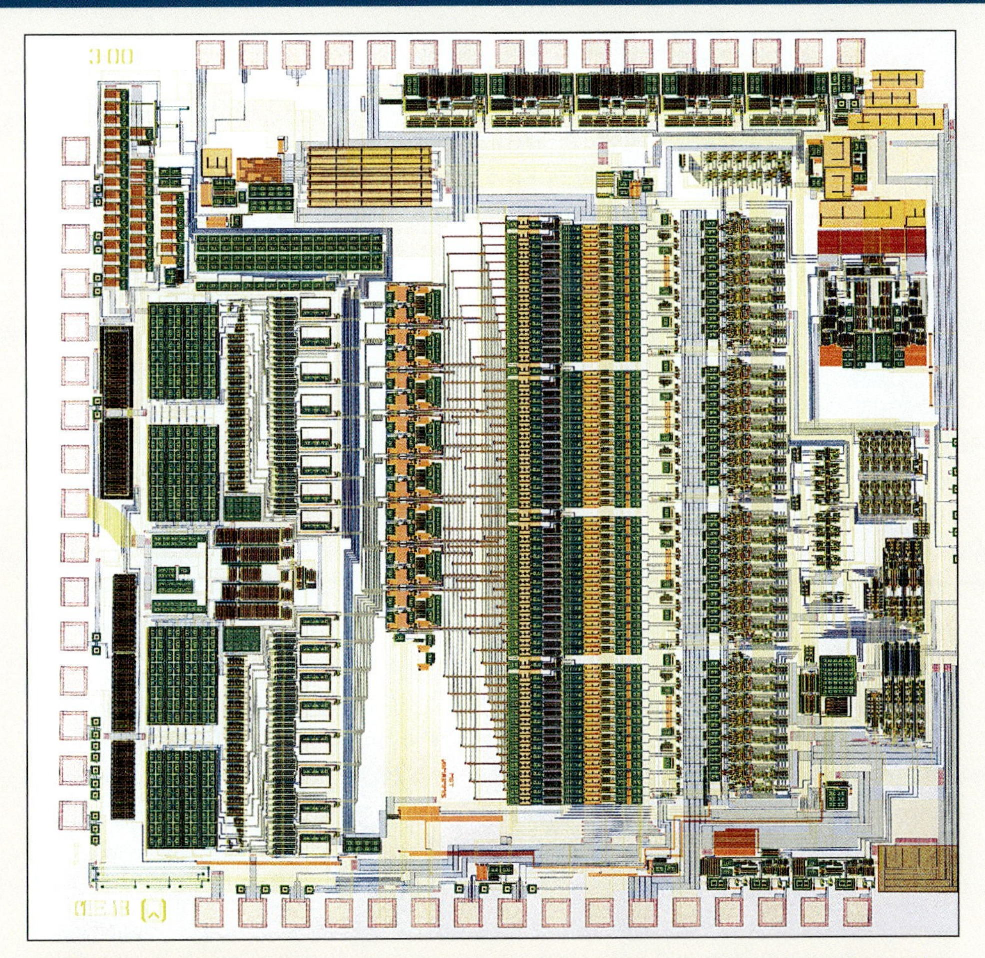

Integrated circuits such as this one have promoted the contemporary information revolution. ***(DOE/FNAL)***

By using boron oxide instead of lime, modern glassmakers produce borosilicate glass, a glass that is highly resistant to heat and serves well as laboratory ware or ovenware. When scientists substitute lead oxide for lime, they create lead glass, a type of glass that is highly refractive of light. Artisans skillfully shape lead glass into exquisite lead crystal art pieces and table crystal. Physicists use special ingredients such as neodymium to create the glasses used in powerful laser systems. Stained glass windows rely upon various colored glasses for their inspiring translucence. Cobalt

SILICON-BASED SOLAR CELLS

Scientists define *solar photovoltaic conversion* as the direct conversion of sunlight (solar energy) into electrical energy by means of the photovoltaic effect. *Solar cell* is the name engineers have given to a single photovoltaic (PV) converter cell. They call the various combinations of cells designed to increase the electric power output either a solar array or a solar panel.

The typical silicon-based solar cell is made up of a combination of n-type (negative) and p-type (positive) semiconductor materials. When this combination of materials is exposed to sunlight, some of the incident electromagnetic radiation removes bound electrons from atoms in the semiconductor material, thereby producing free electrons. Light shining upon pure crystalline silicon may free electrons within the substance's crystal lattice. However, for these free electrons to perform a useful function, such as providing a flow of direct electric current, they must be separated and then directed into an appropriate electrical circuit. To separate the internal electric charges created by the incident light, the silicon solar cell needs an internal (built-in) electric field.

Engineers create this electric field within a solar cell by sandwiching two separate semiconductor materials together. P-type semiconductors have an abundance of positively charged holes, while n-type semiconductors have an abundance of negatively charged electrons. When n- and p-type silicon semi-

compounds mixed with the basic ingredients for glass result in a translucent blue glass, chromium compounds produce green glass, and selenium compounds yield vivid red colors.

CERAMICS

Ceramics are inorganic, nonmetallic substances that usually exhibit high levels of strength at elevated temperatures. The use of ceramics extends back into prehistoric times. Early peoples learned how to work and heat clay to create the first useful ceramics—pottery and bricks. Modern scientists have developed advanced ceramic materials suitable for many specialized applications, including as thermal protection systems for aerospace vehicles.

conductors come into contact with each other, excess electrons move from the n-type side to the p-type side of the interface. The result is a buildup of positive charge along the n-type side of the interface and a buildup of negative charge along the p-type side of the interface.

Because of the flow of electrons and holes, the two semiconductor materials behave like a battery, creating an electric field at the surface where they meet. Engineers refer to this interface as the p-n junction. The electric field causes the electrons to move from the semiconductor toward the negative surface, making them available for flow through an external electric circuit. At the same time, the holes travel in the opposite direction, toward the positive surface, where they await the arrival of incoming electrons.

If electric contacts are made with the n- and p-type materials and these contacts are connected by a conductor through an external load, the free electrons will flow from the n-type material to the p-type material. Upon reaching the p-type semiconductor material, the free electrons will enter existing holes and once again become bound electrons. The flow of free electrons through the external load (conductor) is a direct electric current that will continue as long as more free electrons and holes are being created by exposure of the solar cell to sunlight. This is the general principle of solar photovoltaic conversion, an important process by which the abundant element silicon may help relieve much of humankind's electric energy needs in the future.

As part of the technical ascendancy of humankind, many ancient peoples learned how to use fire to process and transform the various natural materials (rocks and soil) they found in their local environments. One such natural material, clay, proved especially important. Geologists assign the name *clay* to a number of fine-grained earth materials that typically contain silica, alumina, and water. Despite common misperceptions, clay is not a single mineral, but a class of very fine earthen materials that vary widely in composition.

Geologists find it helpful to divide clay into six basic categories. These are common clay, ball clay, fire clay, fuller's earth, kaolin, and bentonite. People use common clay in the manufacture of basic construction materials such as bricks and cement. Ball clays are good quality clays that are used in pottery making. The term *fire clay* applies to all clays (except ball

Artist's rendering showing Native Americans in the southern coastal plain of North Carolina making pottery during the Early Woodland Period (ca. 1000 to 300 B.C.E.) *(National Park Service)*

clays and bentonite) that are used to make items resistant to extreme heat. Fire clay products are often referred to as refractory products. Fuller's earth, composed of the mineral palygorskite, finds its primary use as an absorbent material. Kaolin is soft white plastic clay that consists primarily of the alumino-silicate mineral kaolinite. This fairly common type of clay has industrial applications in the manufacture of paper and in the production of whiteware. The term *whiteware* refers to ceramic products such as tableware, sanitary ware (that is, commodes and other toilet fixtures), and tiles for walls and floors. Finally, bentonite is a highly absorbent form of clay that consists primarily of weathered volcanic ash. One of the common applications of bentonite is pet litter.

Clay is found all over the world. One of the major technology milestones for any rising ancient civilization was the moment its people learned how to use clay to make bricks and pottery. For survival, prehistoric peoples developed storage containers for food and drink. Nomadic cultures generally employed fairly lightweight storage containers usually consisting of basket materials, carved out plants such as gourds, or animal skins. Once people settled down, became farmers, and began to construct permanent habitats, the weight of food and beverage storage containers was no longer of great importance, so artisans began to make household storage items out of clay.

Almost everywhere in the world that clay is found, early agricultural societies developed brick making and pottery making skills. The firing of bricks and pottery to improve their properties represented the birth of the ceramics industry. In some ancient societies, the majority of the pottery was manufactured by skilled tradespersons; in other societies, pottery making became a cottage industry practiced by individuals in each household. Today, archaeologists depend on pottery fragments to learn a great deal about these ancient cultures. Decorated pottery fragments provide archaeologists information about the specific materials early peoples used, the level of each society's technical skills, and their cultural practices and even religious ceremonies.

By accommodating food and beverage storage, pottery helped sustain early agricultural societies during nongrowing seasons. Some pottery, such as the long-necked ceramic jar called an amphora, supported not only household storage but also commerce throughout the Greco-Roman world. Some amphorae were highly decorated and accommodated a wide variety of ceremonial and social functions. Other amphorae were plain, undecorated, and inexpensive. These large, mass-produced Greek and later Roman ceramic jars served as one-way, expendable shipping containers that carried cargoes of wine, olive oil, grain, fish, and similar commodities throughout the ancient Mediterranean world.

The great majority of ceramic materials can resist heat and chemical attack but are brittle and do not conduct electricity well. The porcelain used in modern toilet bowls and other bathroom facilities is an example of a commonly encountered sanitary application of specialized ceramic materials. Materials such as boron carbide (B_4C), silicon carbide (SiC), and alumina (aluminum oxide [Al_2O_3]) are examples of modern ceramic materials sometimes collectively referred to by engineers as *technical ceramics*. One major difficulty with most ceramic materials is that they are brittle and hard. When these materials fail, they fail catastrophically, often breaking with little warning in an irreparable manner.

Industrial engineers sometimes divide the production of ceramics into two broad categories: traditional ceramics and modern ceramics. Traditional ceramic products include whitewares (such as dishes, tiles, and sanitary ware), construction materials (such as bricks and concrete), clay products (such as pottery and sewage treatment and transport materials), and refractory materials (such as furnace bricks, crucibles, and molds). Modern ceramic products include aerospace components (such as the silicon nitride [Si_3N_4] bearings for use in the main rocket engines of NASA's space shuttle), specialized electronic components, and nuclear reactor fuel rods and control rods. Scientists have also developed high-temperature structural ceramics for use as turbine blades, in cutting tools, and in advanced materials processing furnaces. Some modern ceramics are superconductor materials; others exhibit enhanced magnetic properties. Finally, electrical engineers use modern ceramics in a variety of important applications, including insulators, capacitors, semiconductor substrates, piezoelectric devices, superconductors, powerful magnets, and advanced packaging for integrated circuits.

CHAPTER 10

Polymers, Soft Matter, and Composites

People around the world make daily use of a great number of fascinating human-engineered substances, such as plastics and composite materials. While physicists are challenged to explain how various types of soft matter function at the microscopic level, nonscientists go about their daily lives squeezing tubes of toothpaste and hair gel, enjoying whipped cream topping on gelatin desserts, and cleaning dishes in sinks filled with mountains of foamy soap suds. This chapter introduces polymers (especially plastics), soft matter, and composite materials and discusses how each of these special types of solid matter influences daily life.

POLYMERS

Polymers are basically very large molecules, sometimes called macromolecules, that consist of numerous small repeating molecular units called *monomers* that are usually joined together by covalent chemical bonds. Natural polymers include wood, silk, rubber, and wool; human-engineered (or manufactured) polymers include nylon, polyester, and a great variety of other synthetic materials.

How the individual monomers in a polymer are bonded together strongly influences the substance's physical properties. Natural and synthetic rubbers are examples of *elastomers,* amorphous polymers that exhibit well-defined elastic behavior (elasticity). The word *plastic,* when used as a noun, is the popular name that people give to a large family of organic substances (polymers) that can be injected into molds and cast into various shapes, or else extruded and drawn into thin filaments and sheets.

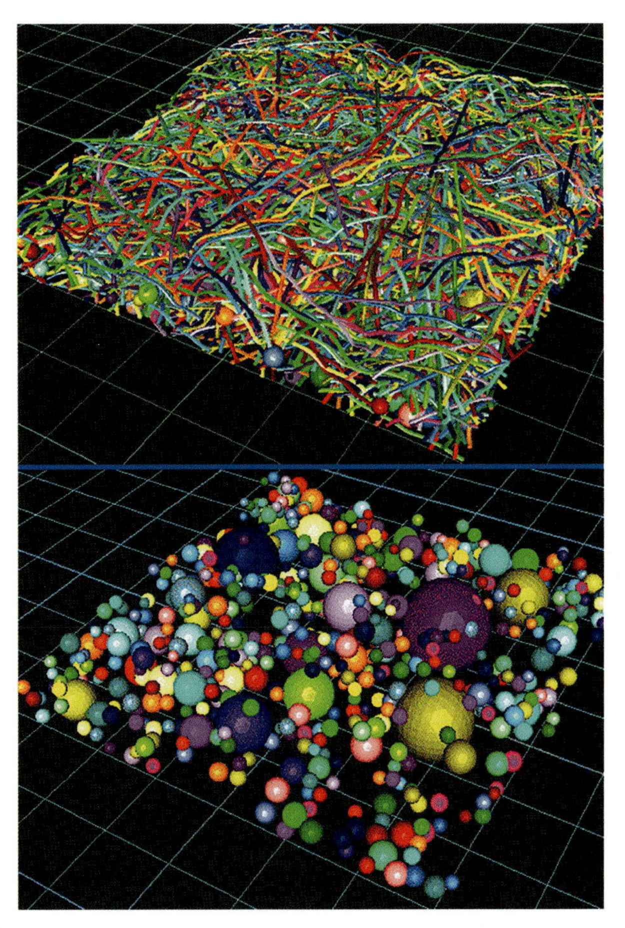

Computer models of fiber-reinforced polymer composites used in automotive structures. Top image shows deposited fibers with void space shown as spheres; bottom image shows void space only. *(DOE/ORNL)*

Scientists note that some polymers have highly crystalline structures, indicating that their long molecules form neat, orderly lines (fibers) that possess great strength. Other polymers have long molecules that are quite randomly oriented, resulting in a rather shapeless (or amorphous) substance that resembles a tangled lump of spaghetti. From a materials science perspective, polymers that are highly crystalline result in good quality synthetic fibers, while polymers that are amorphous in molecular structure tend to produce good elastomers. Some synthetic polymers combine the characteristics of both rigidity (due to the presence of crystalline molecular arrangements) and flexibility (due to the presence of amorphous molecular arrangements). Combined polymers are both interesting from a materials science perspective and useful. Some combined polymers are used to manufacture the specialized fabrics found in exercise clothing and sportswear, such as ski pants and competition swimwear.

Wood is a natural polymer that has been used by humans since antiquity for shelter, furniture, and transportation vehicles, such as carts and ships. Even today, wood remains an extensively employed construction material. Engineers use wood directly in the construction of houses and buildings. They also use wood as a component in a variety of human-engineered composite materials, such as plywood and particle board. (Composite materials are discussed later in the chapter.)

Botanists place trees into two major groups or categories: softwoods and hardwoods. Fir, pine, spruce, and cedar are examples of softwoods. These evergreen trees, known as gymnosperms, have seeds exposed in a conelike structure and retain their leaves throughout the year. Elm, oak, birch, cherry, and maple are examples of hardwoods. These deciduous trees, known as angiosperms, have covered seeds and shed their leaves annually.

The major components of a wood cell are cellulose, lignin, and hemicellulose. About 45 to 50 percent of a tree's solid wood material consists of a crystalline molecular structure called *cellulose,* a natural linear polymer made up of glucose monomers. The glucose monomers share covalent bonds, producing a stiff, crystalline polymer (macromolecule) that is linear and has good tensile strength. Cellulose is the main component of the cell walls of all higher-level plants, including trees. Lignin is a complex organic polymer found within plant cell walls. This polymer plays an important role in transporting water within the plant and makes the walls of a plant rigid, or "woody." Hemicellulose is the amorphous organic polymer found in plant walls along with cellulose. This amorphous, relatively short-chained organic polymer consists of many different types of sugar monomers.

Engineers assess the suitability of various woods for construction applications with the help of the following properties: mechanical strength, density, moisture content, and shrinkage. Green (or freshly harvested) wood usually shrinks as stored-up moisture leaves. To avoid the problems of shrinkage and warping, green wood is often dried in a kiln before use. People have employed wood since prehistoric times. In the past century, however, materials scientists began designing composite materials that take advantage of wood's favorable properties while avoiding some of its disadvantages. Plywood and particle boards are examples of such human-engineered composite materials.

AMAZING WORLD OF PLASTICS

In the late 19th century, the American inventor John Wesley Hyatt (1837–1920) began altering natural organic polymers, especially cellulose from cotton and wood, in an attempt to produce a synthetic substitute for the ivory used to make billiard balls. Using an ethyl alcohol and camphor treatment, he eventually developed a process for softening cellulose nitrate (also known as smokeless gun powder). Hyatt observed that his new synthetic material, which he called *celluloid,* could be fashioned into smooth hard balls. Sometimes regarded as the first human-engineered plastic, Hyatt's celluloid was then made into film for motion pictures; tool, knife, and gun handles; buttons, combs, and brushes; and even shirt collar stiffeners. Celluloid billiard balls also appeared, perhaps sparing a few African elephants from senseless slaughter by ivory hunters. Since celluloid is extremely flammable, this material was rapidly displaced when safer plastic materials began to appear in the marketplace. Today, the only remaining practical use for celluloid is in the manufacture of Ping-Pong balls.

In 1909, the Belgian-American chemist Leo Hendrik Baekeland (1863–1944) announced his invention of a synthetic substance called *Bakelite.* His material was the first plastic made from all synthetic materials and quickly found use in a variety of popular applications. These applications included kitchenware, jewelry, toys, electrical insulation, and as outer cases for electronic equipment such as telephones and radios. The DuPont Company introduced the synthetic thermoplastic material *nylon* in the late 1930s. Nylon proved extremely popular and was fabricated into a wide variety of products ranging from surgical sutures to clothing, screening to fishing line, and tire cords to wire insulation. World War II and the shortage of traditional materials encouraged the development of other new plastic materials.

Scientists and engineers learned how to apply heat and pressure to various organic compounds, especially hydrocarbons, to create plastic materials that performed a wide range of jobs. Plastics now represent an integral material component of civilization. There are two general types of plastics: *thermoplastic* and *thermosetting.* Materials scientists can use heat and pressure to repeatedly soften and remold thermoplastic materials. In contrast, once molded and set in some initial configuration or product, thermosetting materials cannot be reheated and remolded into another useful shape or object.

Some of today's more commonly encountered plastics are briefly mentioned here. Polyethylene is the most popular plastic in the world. It is used for plastic bags, bottles, toys, and electrical insulation. The thermoplastic polymer polypropylene is used for bottles, luggage, and indoor-outdoor carpeting. Simulated wood furniture, drinking cups, toys, cooler chests, packing materials, and thermal insulation involve the use of the thermoplastic polystyrene. Polyvinyl chloride (PVC) is used for pipes, garden hoses, floor tile, plastic wrap, and simulated leather. Food wrap and seat covers are manufactured with polyvinylidene chloride (PVDC). Acrylic fibers are synthetic fibers made with the polymer polyacrylonitrile (PAN). Common PAN products include yarn, wigs, and paint. The polymer polyvinyl acetate (PVA) goes into the production of adhesives, textile coatings, and paints. Polymethyl methacrylate (PMMA) is used for a shatterproof clear plastic substitute for glass. PMMA is also used in the manufacture of bowling balls. Polyurythane (PUR) represents the most common polymer for the production of plastic foams, adhesives, and spandex elastomers. The human-engineered fluoropolymer polytetrafluoroethylene (PTFE) is more commonly recognized by its DuPont brand name Teflon®. The nonstick characteristics of this synthetic poly-

PROBLEM OF PLASTIC POLLUTION

One of the important properties of many plastics is that they are quite durable and can resist a wide variety of environmental conditions. Unfortunately, these distinct advantages also become definite disadvantages when it is time to discard the particular plastic item. Some environmental protection advocates proclaim that "plastics are forever." While this statement may not be entirely correct, there is no question that once plastics are dumped carelessly into the environment, they generally do not disappear. As a reference point, plastic products make up about 10 percent (by mass) of the solid waste disposed of annually in the United States. While disappearing landfill capacity usually represents a regional problem, especially for highly populated urban areas, improperly disposed of plastic items represent a national and international problem. Discarded plastic items now litter streets, countrysides, highways, national parks, and marine environments. Furthermore, plastic litter can be especially harmful to wildlife and marine life.

There is a huge patch of garbage floating in the middle of the Pacific Ocean. Scientists estimate that the Great Pacific Garbage Patch covers an area of between 262,000 and 524,000 square miles (679,100 to 1,358,260 km^2). It is located about 1,000 miles (1,610 km) north of Hawaii in an ocean region known as the North Pacific Gyre. Scientists postulate that ocean currents from North American and Asia converge in this region and accumulate all the floating trash, mostly plastic items, that was dumped into the sea.

mer revolutionized the cookware industry. These are just a few of the amazing plastics that have transformed modern living. New synthetic polymers are being developed almost daily.

SOFT MATTER

Scientists define *soft matter* as matter that exhibits a combination of solid and fluid properties on the macroscopic scale, while having its structure and dynamic behavior at the microscopic scale dominated by thermal fluctuations or thermal stresses. Whipped cream, jelly, gelatin, custard, hair gels, soap bubbles, and plastic foam are all examples of soft matter. One common feature that the different categories of soft

Plastic materials have greatly improved modern life and will undoubtedly remain an integral part of humans' global civilization throughout this century, but plastic pollution can and should be avoided. Proper disposal of all plastic products, recycling, and the expanded use of biodegradable and photodegradable plastic materials would minimize plastic pollution. With heightened environmental awareness, plastics represent a milestone material category (like metals)—a family of synthetic solid substances that promote the continued technical ascendency of humankind.

Although plastics have vastly improved modern life, carelessly discarded plastic materials can create marine debris that can prove fatal to marine mammals, such as this seal. ***(NOAA)***

Scientists in one exciting area within materials science, called *green chemistry,* are exploring the use of biodegradable starch-polyester compositions for the manufacture of one-time-use plastic items, such as knives, forks, spoons, bags, and wraps. These commonly used items are now produced with petroleum-based plastics. Such "thinking green" strategies have the potential to minimize plastic pollution and also reduce consumption of valuable, nonrenewable hydrocarbons.

matter appear to share is that thermal fluctuations take place at energy levels associated with room temperature conditions—nominally about 68°F (20°C [293 K]). Scientists generally include the following types of substances under the broad category of soft matter: colloids, foams, gels, granular materials, liquid crystals, and polymers. All of these materials are fascinating topics, but this section limits the discussions to foam.

Scientists generally define *foam* as a substance that forms by trapping gas bubbles in a liquid or solid matrix. Typical foams have a great variety of bubble sizes and are randomly structured. Shaving cream and soap suds represent aqueous foams that consist of about 95 percent gas and about 5 percent liquid. Despite being almost completely made of air, the sudsy foam created by a mixture of water and liquid dish detergent in a kitchen sink often behaves like a springy solid, while shaving cream dispensed from an aerosol can behaves like a smooth, flexible solid that readily adheres to the side of a person's face. Scientists suggest that the gas subdivides the liquid in aqueous foams into a matrix of tiny bubbles. "Really good" foams usually contain complex molecules that toughen the walls of the bubbles and make the aqueous foams appear stiff. The molecules found in milk fat, for example, give whipped cream a thick, stiff texture.

Whipped cream and the filling for pumpkin pie are two familiar examples of soft matter. *(NASA)*

One of the most important industrial applications of engineered foams is in firefighting. Fire-suppressant foams are especially useful in fighting hydrocarbon fires that occur aboard ships, at commercial airports and military air bases, and in severe vehicle accidents on highways. Some foam systems cover the fuel, hydrocarbon, and smother the fire by preventing oxygen in the air from reaching it. Other foams are designed to suppress the release of flammable vapors that can mix with the air. Still other firefighting foams contain a sufficient water content to cool the hydrocarbon fuel, preventing a conflagration. Well-designed foams, with the appropriate stiffness, bubble size, and dispersion properties are an integral part of modern firefighting systems.

CATCHING COMET DUST WITH AEROGEL

NASA's *Stardust* spacecraft was the first American deep space mission dedicated solely to the exploration of a comet and the first mission designed to return extraterrestrial material from outside the orbit of the Moon. Key to the mission's success was the use of an amazing solid material, aerogel.

A dramatic demonstration of the extraordinary thermal insulating characteristics of space-age aerogel material *(NASA)*

Following its launch from Cape Canaveral, Florida, on February 7, 1999, the spacecraft traveled through interplanetary space and successfully flew by the nucleus of Comet Wild 2 on January 2, 2004. When *Stardust* flew past the comet's nucleus, it did so at an approximate relative velocity of 3.8 miles per second (6.1 km/s). At closest approach during this encounter, the spacecraft came within 155 miles (250 km) of the comet's nucleus. As the spacecraft flew through the dense cloud of gas and dust surrounding the comet's icy nucleus, a special collection grid filled with aerogel, a novel spongelike material that is more than 99 percent empty space, gently captured particle samples. The collection grid was then automatically stowed in a sample return capsule, which later detached itself from the spacecraft and returned to Earth successfully on January 15, 2006.

Since that time, scientists from around the world have been analyzing the captured comet samples. On August 17, 2009, NASA scientists announced that they had discovered glycine, a fundamental building block of life, in the comet samples collected by the *Stardust* spacecraft. Mission scientists were elated by the discovery of glycine and suggested these findings support a theory in astrobiology that some of life's ingredients formed in space and were delivered to Earth early in its history by meteorite and comet impacts.

To enable this mission to successfully collect high-speed particles without damaging them, aerospace engineers used aerogel as the capture material. For

(continues)

(continued)

comparison, aerogel is about 1,000 times less dense than glass and just a little denser than air. When a comet dust particle hit the aerogel, it buried itself in the material, creating a distinctive track that extended in length up to 200 times the particle's own diameter. The high-speed particle slowed down and came to a gradual stop, much like a rifle bullet that is fired into a stack of pillows or a thick bale of cotton. Mission scientists located many interesting particles by following the telltale impact tracks they left in the aerogel. Each particle was then carefully removed and submitted to the scientific community for detailed investigation.

Aerogel is not like conventional foams. It is a special porous material with many individual microscopic features that are well interlinked. This exotic solid substance is strong and has an extremely low thermal conductivity. As shown in the accompanying picture, a thin slab of aerogel is an incredibly good thermal insulator. The silica aerogel used onboard ***Stardust*** was fabricated at NASA's Jet Propulsion Laboratory in California and had the well-controlled properties and purity necessary for successfully capturing high-speed particles.

National security experts are exploring the development of specialized foams to counter the possible use of biological weapons by terrorist groups or rogue nations. Once a germ warfare attack is detected, security personnel might inject such specialized foams into the cracks and crevasses of potentially contaminated areas and kill any dangerous microorganisms hiding there. There is a lot more to foam and soft matter physics than examining the delicious swirl of whipped cream that floats on top of a refreshing cup of iced café mocha.

COMPOSITE MATERIALS

Scientists define a *composite material* as a human-engineered combination of two or more distinct materials, each of which retains its own distinct properties. The blending produces a new material with properties exceeding those of each individual component material. The use of composite materials (or composites) extends from antiquity to the space age. Early civilizations in the Middle East made bricks out of clay and straw. Roman engineers pioneered the use of the ceramic composite known as concrete. Millennia later, aerospace engineers developed reinforced carbon-carbon (RCC) to protect spacecraft from the searing temperatures experienced during atmospheric reentries.

In designing composites, materials scientists and engineers try to offset the weakness inherent in one material of the composite with the strength of the other material(s). The constituents of composite materials are typically divided into two broad categories: the matrix material and the reinforcing material. The matrix material can be a metal, a ceramic, or a polymer; the reinforcing material can be fibers, sheets, or particles that are easily integrated into the matrix material.

Plywood is one of the most common composite materials. These human-engineered boards consist of alternating layers of thin sheets of wood glued together in such a way that the alternating layers have different grain directions. As a result, the engineered material is stronger than a solid board of natural wood of the same dimensions. Similarly, particle board is a human-engineered composite consisting of chips, sawdust, and sawmill shavings embedded in a plastic resin or similar binder. Particle boards are pressed and extruded to various sizes. They represent a relatively inexpensive family of wood products that serve in a number of applications where strength and appearance are much less significant than cost.

Perhaps the most exciting applications of composite materials involve aerospace activities and space exploration. Protecting the leading edges of a spacecraft from the intense temperatures generated during atmospheric reentry operations challenged aerospace engineers to develop reinforced carbon-carbon (RCC) thermal protection materials. The materials needed to be lightweight and capable of withstanding temperatures approaching 3,000°F (1,650°C [1,923 K]). They began using the composite on reentry vehicles. Later, NASA engineers placed RCC on the wing leading edges, the nose cap, and other selected areas of space shuttle orbiter vehicles. RCC is basically a collection of carbon fibers embedded in layers of resin-impregnated graphite. Production of flight-qualified composite material also required a great deal of elegant materials science processes in the form of heat treatments and chemistry.

CHAPTER 11

Conclusion

Solid matter is very special. Most of the ordinary matter in the universe consists of gases or plasma. People would not be here on the third rock from the Sun if it were not for the tiny amount (less than 2 percent) of elemental materials that accreted out of the primordial solar nebula starting about 5 billion years ago.

All the natural chemical elements found on Earth have their ultimate origins in cosmic events. Since different elements come from different cosmic events, the elements found here on Earth, including those that make life possible, reflect an interesting variety of cosmic phenomena that have taken place in the universe over the past 13.7 billion years. The hydrogen found in water and in hydrocarbon molecules was formed within a few minutes after the big bang, but carbon (the element considered necessary for all terrestrial life) along with all the other light elements (such as calcium and potassium) were formed by nucleosynthesis in the interior of large ancient stars. Heavier elements—those with atomic numbers beyond iron, such as silver, gold, thorium, and uranium—were formed by various neutron capture reactions deep in the interiors of highly evolved stars or during supernovas. Certain light elements, such as lithium, beryllium, and boron, resulted from energetic cosmic ray interactions with the atomic nuclei of hydrogen, helium, or other elements found in interstellar space.

When ancient stars exploded, they expelled the elementally enriched stardust that eventually combined with interstellar hydrogen and helium.

The resulting elementally enriched interstellar gas then became available to create a new generation of stars and, for many of these next generation stars, a family of companion planets. About 5 billion years ago, humans' solar system, including planet Earth, started forming from one such elementally enriched cloud of interstellar gas.

As the solar system formed, most of the mass in the original solar nebula created the Sun, which is basically a gigantic ball of plasma. The leftover matter became everything else—major planets, moons, dwarf planets, asteroids, and comets. The bulk of the available planetary mass went to forming the gaseous giant outer planets (Jupiter, Saturn, Uranus, and Neptune). The very small amount of matter that lingered in the inner solar system became Mercury, Venus, Earth, Mars, and the Moon.

The rocky materials that formed Earth eventually cooled and made its crust sufficiently firm to allow life to emerge from ancient seas and occupy the land. Rocks are truly minor miracles, for without them Earth would not exist and neither would people. Biblical scholars suggest that the ancient Hebrew word *adama* means "earth" or "out of the earth." Is this a literary coincidence or a profound prescientific understanding that the ascendancy of man is linked to the skillful use of Earth's solid matter?

One interesting solid material is clay, the substance that allowed early peoples to make bricks, wall their cities for protection, and start the first civilizations. Clay pottery also played a major role in the establishment of early societies during the Neolithic Revolution. Before the first farmers arose, nomadic prehistoric peoples probably used bags made of animal hides to store water and food. Once people began to stay in one place, they discovered how to make pottery, a development that made the storage of food and drink much easier. As the first societies emerged, larger pieces of pottery supported the shipment of food and liquids, including olive oil and wine, throughout the ancient world.

Nevertheless, without metals, humans would have remained essentially New Stone Age (Neolithic) farmers. Metals are the forms of solid matter that advanced civilization and enabled the rise of technology. It was the use of copper and bronze in ancient times that lifted early civilizations beyond the relativity simple agrarian economy of the Neolithic period to technology-based societies enhanced by the use of metal tools and objects. Precious metals, minted as money, helped move economic development beyond a primitive barter system. Without other metals, such as iron and steel, steam engines and the First Industrial Revolution would not have occurred. The improved scientific understanding of matter, including metals, enabled the development of electricity and information technology. The electric generator and the telephone and telegraph promoted the Second Industrial Revolution.

Without metals, human beings would have remained locked in resource-limited, localized, agrarian-based societies. From skyscrapers to suspension bridges, industrial machinery to ships, and paper clips to automobile frames, steel in all its various compositions serves as the metallic backbone of modern civilization. Iron is the world's most extensively used metal, followed by aluminum, and then a host of other metallic elements, but the skilled manipulation of solid matter did not stop with sophisticated metal alloys.

Starting in the mid-20th century, bits of sand embedded with selected rare earth materials and metals empowered a massive, interconnected global economy based on the continuous flow of information and digital money. In the 21st century, as humans learn to efficiently manipulate matter at the atomic level through nanotechnology, even more exciting new applications of solid matter will influence the trajectory of civilization. Nanotechnology represents an extremely important plateau in human history—somewhat analogous in its overall potential impact to the discovery and use of fire by prehistoric people. What is clearly different in this case is the timescale within which the benefits and risks associated with future discoveries can influence the pathway followed by civilization.

Sweeping technical changes in matter manipulation and materials science can occur in periods as short as decades, or possibly even years. Such radical changes in the manipulation of solid matter will undoubtedly be accompanied by inevitable stresses on existing environmental, social, and economic infrastructures. According to the second law of thermodynamics, nature provides no free lunch. The price of harvesting huge benefits from this anticipated era of rapid technical progress in materials science is heightened social and ethical vigilance.

Any futuristic vision of a golden age of material prosperity must be tempered by the specter of metal-based oblivion. In the second half of the 20th century, humans learned how to manipulate certain types of solid matter, such as small pieces of plutonium, so as to initiate powerful nuclear fission and later thermonuclear explosions. Up to that point in history, metals generally signaled the ascendancy of humans, but with the arrival of the nuclear weapon, the advanced use of metals and other materials represented potential "self-destruction in a can." The human race must apply the latest developments in information technology to acquire an effective, collective ability to think and act responsibly, for only then can people properly manage their ever more powerful technical abilities, including the ability to manipulate solid matter. The human race now faces a great choice. Materials science can open up the universe or else lead to oblivion!

Appendix

Scientists correlate the properties of the elements portrayed in the periodic table with their electron configurations. Since, in a neutral atom, the number of electrons equals the number of protons, they arrange the elements in order of their increasing atomic number (Z). The modern periodic table has seven horizontal rows (called periods) and 18 vertical columns (called groups). The properties of the elements in a particular row vary across it, providing the concept of periodicity.

There are several versions of the periodic table used in modern science. The International Union of Pure and Applied Chemistry (IUPAC) recommends labeling the vertical columns from 1 to 18, starting with hydrogen (H) as the top of group 1 and ending with helium (He) as the top of group 18. The IUPAC further recommends labeling the periods (rows) from 1 to 7. Hydrogen (H) and helium (He) are the only two elements found in

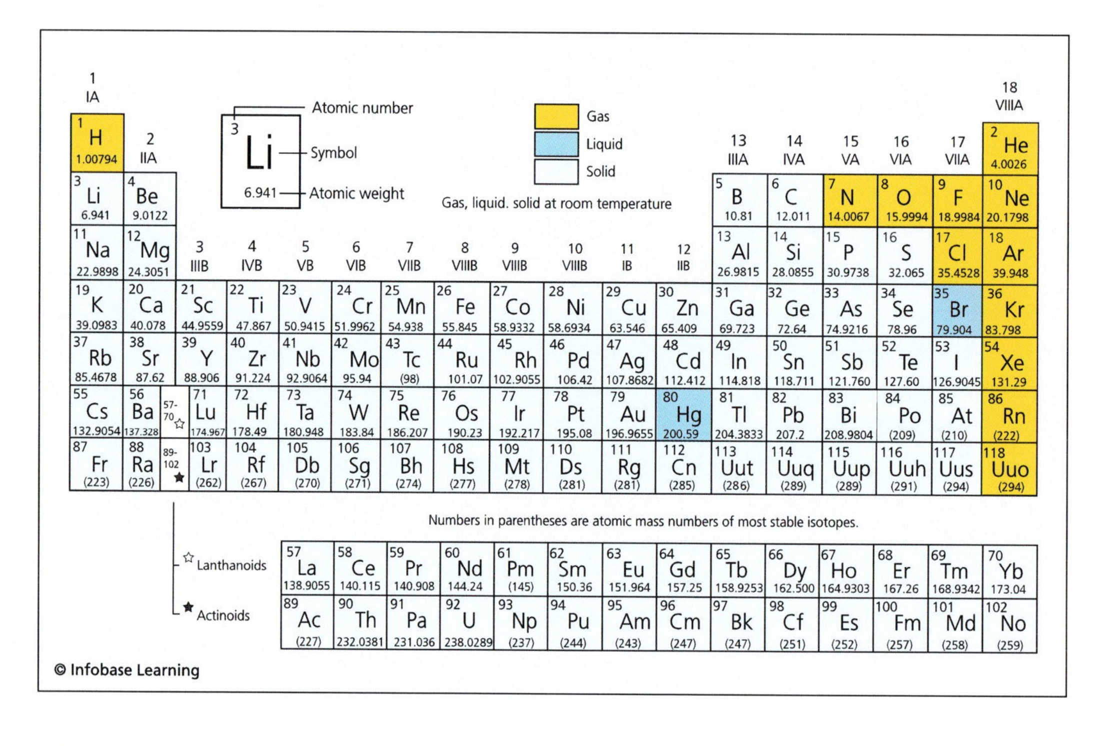

period (row) 1. Period 7 starts with francium (Fr) and includes the actinide series as well as the transactinides (very short-lived, human-made, superheavy elements).

The row (or period) in which an element appears in the periodic table tells scientists how many electron shells an atom of that particular element possesses. The column (or group) lets scientists know how many electrons to expect in an element's outermost electron shell. Scientists call an electron residing in an atom's outermost shell a valence electron. Chemists have learned that it is these valence electrons that determine the chemistry of a particular element. The periodic table is structured such that all the elements in the same column (group) have the same number of valence electrons. The elements that appear in a particular column (group) display similar chemistry.

ELEMENTS LISTED BY ATOMIC NUMBER

1	H	Hydrogen	18	Ar	Argon
2	He	Helium	19	K	Potassium
3	Li	Lithium	20	Ca	Calcium
4	Be	Beryllium	21	Sc	Scandium
5	B	Boron	22	Ti	Titanium
6	C	Carbon	23	V	Vanadium
7	N	Nitrogen	24	Cr	Chromium
8	O	Oxygen	25	Mn	Manganese
9	F	Fluorine	26	Fe	Iron
10	Ne	Neon	27	Co	Cobalt
11	Na	Sodium	28	Ni	Nickel
12	Mg	Magnesium	29	Cu	Copper
13	Al	Aluminum	30	Zn	Zinc
14	Si	Silicon	31	Ga	Gallium
15	P	Phosphorus	32	Ge	Germanium
16	S	Sulfur	33	As	Arsenic
17	Cl	Chlorine	34	Se	Selenium

(continues)

ELEMENTS LISTED BY ATOMIC NUMBER *(continued)*

35	Br	Bromine	63	Eu	Europium
36	Kr	Krypton	64	Gd	Gadolinium
37	Rb	Rubidium	65	Tb	Terbium
38	Sr	Strontium	66	Dy	Dysprosium
39	Y	Yttrium	67	Ho	Holmium
40	Zr	Zirconium	68	Er	Erbium
41	Nb	Niobium	69	Tm	Thulium
42	Mo	Molybdenum	70	Yb	Ytterbium
43	Tc	Technetium	71	Lu	Lutetium
44	Ru	Ruthenium	72	Hf	Hafnium
45	Rh	Rhodium	73	Ta	Tantalum
46	Pd	Palladium	74	W	Tungsten
47	Ag	Silver	75	Re	Rhenium
48	Cd	Cadmium	76	Os	Osmium
49	In	Indium	77	Ir	Iridium
50	Sn	Tin	78	Pt	Platinum
51	Sb	Antimony	79	Au	Gold
52	Te	Tellurium	80	Hg	Mercury
53	I	Iodine	81	Tl	Thallium
54	Xe	Xenon	82	Pb	Lead
55	Cs	Cesium	83	Bi	Bismuth
56	Ba	Barium	84	Po	Polonium
57	La	Lanthanum	85	At	Astatine
58	Ce	Cerium	86	Rn	Radon
59	Pr	Praseodymium	87	Fr	Francium
60	Nd	Neodymium	88	Ra	Radium
61	Pm	Promethium	89	Ac	Actinium
62	Sm	Samarium	90	Th	Thorium

91	Pa	Protactinium
92	U	Uranium
93	Np	Neptunium
94	Pu	Plutonium
95	Am	Americium
96	Cm	Curium
97	Bk	Berkelium
98	Cf	Californium
99	Es	Einsteinium
100	Fm	Fermium
101	Md	Mendelevium
102	No	Nobelium
103	Lr	Lawrencium
104	Rf	Rutherfordium

105	Db	Dubnium
106	Sg	Seaborgium
107	Bh	Bohrium
108	Hs	Hassium
109	Mt	Meitnerium
110	Ds	Darmstadtium
111	Rg	Roentgenium
112	Cn	Copernicum
113	Uut	Ununtrium
114	Uuq	Ununquadium
115	Uup	Ununpentium
116	Uuh	Ununhexium
117	Uus	Ununseptium
118	Uuo	Ununoctium

Chronology

Civilization is essentially the story of the human mind understanding and gaining control over matter. The chronology presents some of the major milestones, scientific breakthroughs, and technical developments that formed the modern understanding of matter. Note that dates prior to 1543 are approximate.

13.7 BILLION YEARS AGO Big bang event starts the universe.

13.3 BILLION YEARS AGO The first stars form and begin to shine intensely.

4.5 BILLION YEARS AGO Earth forms within the primordial solar nebula.

3.6 BILLION YEARS AGO Life (simple microorganisms) appears in Earth's oceans.

2,000,000–100,000 B.C.E. Early hunters of the Lower Paleolithic learn to use simple stone tools, such as handheld axes.

100,000–40,000 B.C.E. Neanderthal man of Middle Paleolithic lives in caves, controls fire, and uses improved stone tools for hunting.

40,000–10,000 B.C.E. During the Upper Paleolithic, Cro-Magnon man displaces Neanderthal man. Cro-Magnon people develop more organized hunting and fishing activities using improved stone tools and weapons.

8000–3500 B.C.E. Neolithic Revolution takes place in the ancient Middle East as people shift their dependence for subsistence from hunting and gathering to crop cultivation and animal domestication.

3500–1200 B.C.E. Bronze Age occurs in the ancient Middle East, when metalworking artisans start using bronze (a copper and tin alloy) to make weapons and tools.

1200–600 B.C.E. People in the ancient Middle East enter the Iron Age. Eventually, the best weapons and tools are made of steel, an alloy of iron and varying amounts of carbon. The improved metal tools and weapons spread to Greece and later to Rome.

1000 B.C.E. . . . . By this time, people in various ancient civilizations have discovered and are using the following chemical elements (in alphabetical order): carbon (C), copper (Cu), gold (Au), iron (Fe), lead (Pb), mercury (Hg), silver (Ag), sulfur (S), tin (Sn), and zinc (Zn).

650 B.C.E. . . . . Kingdom of Lydia introduces officially minted gold and silver coins.

600 B.C.E. . . . . Early Greek philosopher Thales of Miletus postulates that all substances come from water and would eventually turn back into water.

450 B.C.E. . . . . Greek philosopher Empedocles proposes that all matter is made up of four basic elements (earth, air, water, and fire) that periodically combine and separate under the influence of two opposing forces (love and strife).

430 B.C.E. . . . . Greek philosopher Democritus proposes that all things consist of changeless, indivisible, tiny pieces of matter called *atoms.*

250 B.C.E. . . . . Archimedes of Syracuse designs an endless screw, later called the Archimedes screw. People use the fluid-moving device to remove water from the holds of sailing ships and to irrigate arid fields.

300 C.E. . . . . Greek alchemist Zosimos of Panoplis writes the oldest known work describing alchemy.

850 . . . . The Chinese use gunpowder for festive fireworks. It is a mixture of sulfur (S), charcoal (C), and potassium nitrate (KNO_3).

1247 . . . . British monk Roger Bacon writes the formula for gunpowder in his encyclopedic work *Opus Majus.*

1250 . . . . German theologian and natural philosopher Albertus Magnus isolates the element arsenic (As).

1439 . . . . Johannes Gutenberg successfully incorporates movable metal type in his mechanical printing press.

His revolutionary approach to printing depends on a durable, hard metal alloy called type metal, which consists of a mixture of lead (Pb), tin (Sn), and antimony (Sb).

1543 .... Start of the Scientific Revolution. Polish astronomer Nicholas Copernicus promotes heliocentric (Sun-centered) cosmology with his deathbed publication of *On the Revolutions of Celestial Orbs.*

1638 .... Italian scientist Galileo Galilei publishes extensive work on solid mechanics, including uniform acceleration, free fall, and projectile motion.

1643 .... Italian physicist Evangelista Torricelli designs the first mercury barometer and then records the daily variation of atmospheric pressure.

1661 .... Irish-British scientist Robert Boyle publishes *The Sceptical Chymist,* in which he abandons the four classical Greek elements (earth, air, water, and fire) and questions how alchemists determine what substances are elements.

1665 .... British scientist Robert Hooke publishes *Micrographia,* in which he describes pioneering applications of the optical microscope in chemistry, botany, and other scientific fields.

1667 .... The work of German alchemist Johann Joachim Becher forms the basis of the phlogiston theory of heat.

1669 .... German alchemist Hennig Brand discovers the element phosphorous (P).

1678 .... Robert Hooke studies the action of springs and reports that the extension (or compression) of an elastic material takes place in direct proportion to the force exerted on the material.

1687 .... British physicist Sir Isaac Newton publishes *The Principia.* His work provides the mathematical foundations for understanding (from a classical physics perspective) the motion of almost everything in the physical universe.

1738 .... Swiss mathematician Daniel Bernoulli publishes *Hydrodynamica*. In this seminal work, he identifies the relationships between density, pressure, and velocity in flowing fluids.

1748 .... While conducting experiments with electricity, American statesman and scientist Benjamin Franklin coins the term *battery*.

1754.................. Scottish chemist Joseph Black discovers a new gaseous substance, which he calls "fixed air." Other scientists later identify it as carbon dioxide (CO_2).

1764 .... Scottish engineer James Watt greatly improves the Newcomen steam engine. Watt steam engines power the First Industrial Revolution.

1772 .... Scottish physician and chemist Daniel Rutherford isolates a new colorless gaseous substance, calling it "noxious air." Other scientists soon refer to the new gas as nitrogen (N_2).

1785 .... French scientist Charles-Augustin de Coulomb performs experiments that lead to the important law of electrostatics, later known as Coulomb's law.

1789 .... French chemist Antoine-Laurent Lavoisier publishes *Treatise of Elementary Chemistry*, the first modern textbook on chemistry. Lavoisier also promotes the caloric theory of heat.

1800 .... Italian physicist Count Alessandro Volta invents the voltaic pile. His device is the forerunner of the modern electric battery.

1803 .... British schoolteacher and chemist John Dalton revives the atomic theory of matter. From his experiments, he concludes that all matter consists of combinations of atoms and that all the atoms of a particular element are identical.

1807 .... British chemist Sir Humphry Davy discovers the element potassium (K) while experimenting with caustic potash (KOH). Potassium is the first metal isolated by the process of electrolysis.

1811.................... Italian physicist Amedeo Avogadro proposes that equal volumes of different gases under the same conditions of pressure and temperature contain the same number of molecules. Scientists call this important hypothesis Avogadro's law.

1820 ....Danish physicist Hans Christian Ørsted discovers a relationship between magnetism and electricity.

1824...................French military engineer Sadi Carnot publishes *Reflections on the Motive Power of Fire.* Despite the use of caloric theory, his work correctly identifies the general thermodynamic principles that govern the operation and efficiency of all heat engines.

1826 .... French scientist André-Marie Ampère experimentally formulates the relationship between electricity and magnetism.

1827 .... Experiments performed by German physicist George Simon Ohm indicate a fundamental relationship among voltage, current, and resistance.

1828 .... Swedish chemist Jöns Jacob Berzelius discovers the element thorium (Th).

1831 .... British experimental scientist Michael Faraday discovers the principle of electromagnetic induction. This principle is the basis for the electric dynamo.

Independent of Faraday, the American physicist Joseph Henry publishes a paper describing the electric motor (essentially a reverse dynamo).

1841 .... German physicist and physician Julius Robert von Mayer states the conservation of energy principle, namely that energy can neither be created nor destroyed.

1847 .... British physicist James Prescott Joule experimentally determines the mechanical equivalent of heat. Joule's work is a major step in developing the modern science of thermodynamics.

1866 .... Swedish scientist-industrialist Alfred Nobel finds a way to stabilize nitroglycerin and calls the new chemical explosive mixture dynamite.

1869 .... Russian chemist Dmitri Mendeleev introduces a periodic listing of the 63 known chemical elements in *Principles of Chemistry.* His periodic table includes gaps for elements predicted but not yet discovered.

American printer John W. Hyatt formulates celluloid, a flammable thermoplastic material made from a mixture of cellulose nitrate, alcohol, and camphor.

1873 .... Scottish mathematician and theoretical physicist James Clerk Maxwell publishes *Treatise on Electricity and Magnetism.*

1876 .... American physicist and chemist Josiah Willard Gibbs publishes *On the Equilibrium of Heterogeneous Substances.* This compendium forms the theoretical foundation of physical chemistry.

1884 .... Swedish chemist Svante Arrhenius proposes that electrolytes split or dissociate into electrically opposite positive and negative ions.

1888 .... German physicist Heinrich Rudolf Hertz produces and detects radio waves.

1895 .... German physicist Wilhelm Conrad Roentgen discovers X-rays.

1896 .... While investigating the properties of uranium salt, French physicist Antoine-Henri Becquerel discovers radioactivity.

1897 .... British physicist Sir Joseph John Thomson performs experiments that demonstrate the existence of the electron—the first subatomic particle discovered.

1898 .... French scientists Pierre and (Polish-born) Marie Curie announce the discovery of two new radioactive elements, polonium (Po) and radium (Ra).

1900 .... German physicist Max Planck postulates that blackbodies radiate energy only in discrete packets (or quanta) rather than continuously. His hypothesis marks the birth of quantum theory.

1903 .... New Zealand–born British physicist Baron (Ernest) Rutherford and British radiochemist Frederick Soddy propose the law of radioactive decay.

1904 .... German physicist Ludwig Prandtl revolutionizes fluid mechanics by introducing the concept of the boundary layer and its role in fluid flow.

1905 .... Swiss-German-American physicist Albert Einstein publishes the special theory of relativity, including the famous mass-energy equivalence formula ($E = mc^2$).

1907 .... Belgian-American chemist Leo Baekeland formulates bakelite. This synthetic thermoplastic material ushers in the age of plastics.

1911.................. Baron Ernest Rutherford proposes the concept of the atomic nucleus based on the startling results of an alpha particle–gold foil scattering experiment.

1912 .... German physicist Max von Laue discovers that X-rays are diffracted by crystals.

1913 .... Danish physicist Niels Bohr presents his theoretical model of the hydrogen atom—a brilliant combination of atomic theory with quantum physics.

Frederick Soddy proposes the existence of isotopes.

1914 .... British physicist Henry Moseley measures the characteristic X-ray lines of many chemical elements.

1915 .... Albert Einstein presents his general theory of relativity, which relates gravity to the curvature of space-time.

1919 .... Ernest Rutherford bombards nitrogen (N) nuclei with alpha particles, causing the nitrogen nuclei to transform into oxygen (O) nuclei and to emit protons (hydrogen nuclei).

British physicist Francis Aston uses the newly invented mass spectrograph to identify more than 200 naturally occurring isotopes.

1923 .... American physicist Arthur Holly Compton conducts experiments involving X-ray scattering that demonstrate the particle nature of energetic photons.

1924 .... French physicist Louis-Victor de Broglie proposes the particle-wave duality of matter.

1926 .... Austrian physicist Erwin Schrödinger develops quantum wave mechanics to describe the dual wave-particle nature of matter.

1927 .... German physicist Werner Heisenberg introduces his uncertainty principle.

1929 .... American astronomer Edwin Hubble announces that his observations of distant galaxies suggest an expanding universe.

1932 .... British physicist Sir James Chadwick discovers the neutron.

British physicist Sir John Cockcroft and Irish physicist Ernest Walton use a linear accelerator to bombard lithium (Li) with energetic protons, producing the first artificial disintegration of an atomic nucleus.

American physicist Carl D. Anderson discovers the positron.

1934 .... Italian-American physicist Enrico Fermi proposes a theory of beta decay that includes the neutrino. He also starts to bombard uranium with neutrons and discovers the phenomenon of slow neutrons.

1938 .... German chemists Otto Hahn and Fritz Strassmann bombard uranium with neutrons and detect the presence of lighter elements. Austrian physicist Lise Meitner and Austrian-British physicist Otto Frisch review Hahn's work and conclude in early 1939 that the German chemists had split the atomic nucleus, achieving neutron-induced nuclear fission.

E.I. du Pont de Nemours & Company introduces a new thermoplastic material called nylon.

1941 .... American nuclear scientist Glenn T. Seaborg and his associates use the cyclotron at the University of California, Berkeley, to synthesize plutonium (Pu).

1942 .... Modern nuclear age begins when Enrico Fermi's scientific team at the University of Chicago achieves the first self-sustained, neutron-induced fission chain reaction at Chicago Pile One (CP-1), a uranium-fueled, graphite-moderated atomic pile (reactor).

1945 .... American scientists successfully detonate the world's first nuclear explosion, a plutonium-implosion device code-named Trinity.

1947 .... American physicists John Bardeen, Walter Brattain, and William Shockley invent the transistor.

1952 .... A consortium of 11 founding countries establishes CERN, the European Organization for Nuclear Research, at a site near Geneva, Switzerland.

United States tests the world's first thermonuclear device (hydrogen bomb) at the Enewetak Atoll in the Pacific Ocean. Code-named Ivy Mike, the experimental device produces a yield of 10.4 megatons.

1964 .... German-American physicist Arno Allen Penzias and American physicist Robert Woodrow Wilson detect the cosmic microwave background (CMB).

1967 .... German-American physicist Hans Albrecht Bethe receives the 1967 Nobel Prize in physics for his theory of thermonuclear reactions being responsible for energy generation in stars.

1969 .... On July 20, American astronauts Neil Armstrong and Edwin "Buzz" Aldrin successfully land on the Moon as part of NASA's *Apollo 11* mission.

1972 .... NASA launches the *Pioneer 10* spacecraft. It eventually becomes the first human-made object to leave the solar system on an interstellar trajectory

1985 .... American chemists Robert F. Curl, Jr., and Richard E. Smalley, collaborating with British astronomer Sir Harold W. Kroto, discover the buckyball, an allotrope of pure carbon.

1996 .... Scientists at CERN (near Geneva, Switzerland) announce the creation of antihydrogen, the first human-made antimatter atom.

1998 .... Astrophysicists investigating very distant Type 1A supernovae discover that the universe is expanding at an accelerated rate. Scientists coin the term *dark energy* in their efforts to explain what these observations physically imply.

2001 .... American physicist Eric A. Cornell, German physicist Wolfgang Ketterle, and American physicist Carl E. Wieman share the 2001 Nobel Prize in physics for their fundamental studies of the properties of Bose-Einstein condensates.

2005 .... Scientists at the Lawrence Livermore National Laboratory (LLNL) in California and the Joint Institute for Nuclear Research (JINR) in Dubna, Russia, perform collaborative experiments that establish the existence of super-heavy element 118, provisionally called ununoctium (Uuo).

2008 .... An international team of scientists inaugurates the world's most powerful particle accelerator, the Large Hadron Collider (LHC), located at the CERN laboratory near Geneva, Switzerland.

2009 .... British scientist Charles Kao, American scientist Willard Boyle, and American scientist George Smith share the 2009 Nobel Prize in physics for their pioneering efforts in fiber optics and imaging semiconductor devices, developments that unleashed the information technology revolution.

2010 .... Element 112 is officially named Copernicum (Cn) by the IUPAC in honor of Polish astronomer Nicholas Copernicus (1473–1543), who championed heliocentric cosmology.

Scientists at the Joint Institute for Nuclear Research in Dubna, Russia, announce the synthesis of element 117 (ununseptium [Uus]) in early April.

Glossary

absolute zero the lowest possible temperature; equal to 0 kelvin (K) (–459.67°F, –273.15°C)

acceleration (a) rate at which the velocity of an object changes with time

accelerator device for increasing the velocity and energy of charged elementary particles

acid substance that produces hydrogen ions (H^+) when dissolved in water

actinoid (formerly actinide) series of heavy metallic elements beginning with element 89 (actinium) and continuing through element 103 (lawrencium)

activity measure of the rate at which a material emits nuclear radiations

air overall mixture of gases that make up Earth's atmosphere

alchemy mystical blend of sorcery, religion, and prescientific chemistry practiced in many early societies around the world

alloy solid solution (compound) or homogeneous mixture of two or more elements, at least one of which is an elemental metal

alpha particle (α) positively charged nuclear particle emitted from the nucleus of certain radioisotopes when they undergo decay; consists of two protons and two neutrons bound together

alternating current (AC) electric current that changes direction periodically in a circuit

American customary system of units (also American system) used primarily in the United States; based on the foot (ft), pound-mass (lbm), pound-force (lbf), and second (s). Peculiar to this system is the artificial construct (based on Newton's second law) that one pound-force equals one pound-mass (lbm) at sea level on Earth

ampere (A) SI unit of electric current

anode positive electrode in a battery, fuel cell, or electrolytic cell; oxidation occurs at anode

antimatter matter in which the ordinary nuclear particles are replaced by corresponding antiparticles

Archimedes principle the fluid mechanics rule that states that the buoyant (upward) force exerted on a solid object immersed in a fluid equals the weight of the fluid displaced by the object

atom smallest part of an element, indivisible by chemical means; consists of a dense inner core (nucleus) that contains protons and neutrons and a cloud of orbiting electrons

atomic mass *See* **relative atomic mass**

atomic mass unit (amu) 1/12 mass of carbon's most abundant isotope, namely carbon-12

atomic number (Z) total number of protons in the nucleus of an atom and its positive charge

atomic weight the mass of an atom relative to other atoms. *See also* **relative atomic mass**

battery electrochemical energy storage device that serves as a source of direct current or voltage

becquerel (Bq) SI unit of radioactivity; one disintegration (or spontaneous nuclear transformation) per second. *Compare with* **curie**

beta particle (β) elementary particle emitted from the nucleus during radioactive decay; a negatively charged beta particle is identical to an electron

big bang theory in cosmology concerning the origin of the universe; postulates that about 13.7 billion years ago, an initial singularity experienced a very large explosion that started space and time. Astrophysical observations support this theory and suggest that the universe has been expanding at different rates under the influence of gravity, dark matter, and dark energy

blackbody perfect emitter and perfect absorber of electromagnetic radiation; radiant energy emitted by a blackbody is a function only of the emitting object's absolute temperature

black hole incredibly compact, gravitationally collapsed mass from which nothing can escape

boiling point temperature (at a specified pressure) at which a liquid experiences a change of state into a gas

Bose-Einstein condensate (BEC) state of matter in which extremely cold atoms attain the same quantum state and behave essentially as a large "super atom"

boson general name given to any particle with a spin of an integral number (0, 1, 2, etc.) of quantum units of angular momentum. Carrier particles of all interactions are bosons. *See also* **carrier particle**

brass alloy of copper (Cu) and zinc (Zn)

British thermal unit (Btu) amount of heat needed to raise the temperature of 1 lbm of water 1°F at normal atmospheric pressure; 1 Btu = 1,055 J = 252 cal

bronze alloy of copper (Cu) and tin (Sn)

calorie (cal) quantity of heat; defined as the amount needed to raise one gram of water 1°C at normal atmospheric pressure; 1 cal = 4.1868 J = 0.004 Btu

carbon dioxide (CO_2) colorless, odorless, noncombustible gas present in Earth's atmosphere

Carnot cycle ideal reversible thermodynamic cycle for a theoretical heat engine; represents the best possible thermal efficiency of any heat engine operating between two absolute temperatures (T_1 and T_2)

carrier particle within the standard model, gluons are carrier particles for strong interactions; photons are carrier particles of electromagnetic interactions; and the W and Z bosons are carrier particles for weak interactions. *See also* **standard model**

catalyst substance that changes the rate of a chemical reaction without being consumed or changed by the reaction

cathode negative electrode in a battery, fuel cell, electrolytic cell, or electron (discharge) tube through which a primary stream of electrons enters a system

chain reaction reaction that stimulates its own repetition. *See also* **nuclear chain reaction**

change of state the change of a substance from one physical state to another; the atoms or molecules are structurally rearranged without experiencing a change in composition. Sometimes called change of phase or phase transition

charged particle elementary particle that carries a positive or negative electric charge

chemical bond(s) force(s) that holds atoms together to form stable configurations of molecules

chemical property characteristic of a substance that describes the manner in which the substance will undergo a reaction with another substance, resulting in a change in chemical composition. *Compare with* **physical property**

chemical reaction involves changes in the electron structure surrounding the nucleus of an atom; a dissociation, recombination, or rearrangement of atoms. During a chemical reaction, one or more kinds of matter (called reactants) are transformed into one or several new kinds of matter (called products)

color charge in the standard model, the charge associated with strong interactions. Quarks and gluons have color charge and thus participate in strong interactions. Leptons, photons, W bosons, and Z bosons do not have color charge and consequently do not participate in strong interactions. *See also* **standard model**

combustion chemical reaction (burning or rapid oxidation) between a fuel and oxygen that generates heat and usually light

composite materials human-made materials that combine desirable properties of several materials to achieve an improved substance; includes combinations of metals, ceramics, and plastics with built-in strengthening agents

compound pure substance made up of two or more elements chemically combined in fixed proportions

compressible flow fluid flow in which density changes cannot be neglected

compression condition when an applied external force squeezes the atoms of a material closer together. *Compare* **tension**

concentration for a solution, the quantity of dissolved substance per unit quantity of solvent

condensation change of state process by which a vapor (gas) becomes a liquid. *The opposite of* **evaporation**

conduction (thermal) transport of heat through an object by means of a temperature difference from a region of higher temperature to a region of lower temperature. *Compare with* **convection**

conservation of mass and energy Einstein's special relativity principle stating that energy (E) and mass (m) can neither be created nor destroyed, but are interchangeable in accordance with the equation $E = mc^2$, where c represents the speed of light

convection fundamental form of heat transfer characterized by mass motions within a fluid resulting in the transport and mixing of the properties of that fluid

coulomb (C) SI unit of electric charge; equivalent to quantity of electric charge transported in one second by a current of one ampere

covalent bond the chemical bond created within a molecule when two or more atoms share an electron

creep slow, continuous, permanent deformation of solid material caused by a constant tensile or compressive load that is less than the load necessary for the material to give way (yield) under pressure. *See also* **plastic deformation**

crystal a solid whose atoms are arranged in an orderly manner, forming a distinct, repetitive pattern

curie (Ci) traditional unit of radioactivity equal to 37 billion (37×10^9) disintegrations per second. *Compare with* **becquerel**

current (I) flow of electric charge through a conductor

dark energy a mysterious natural phenomenon or unknown cosmic force thought responsible for the observed acceleration in the rate of expansion of the universe. Astronomical observations suggest dark energy makes up about 72 percent of the universe

dark matter (nonbaryonic matter) exotic form of matter that emits very little or no electromagnetic radiation. It experiences no measurable interaction with ordinary (baryonic) matter but somehow accounts for the observed structure of the universe. It makes up about 23 percent of the content of the universe, while ordinary matter makes up less than 5 percent

density (ρ) mass of a substance per unit volume at a specified temperature

deposition direct transition of a material from the gaseous (vapor) state to the solid state without passing through the liquid phase. *Compare with* **sublimation**

dipole magnet any magnet with one north and one south pole

direct current (DC) electric current that always flows in the same direction through a circuit

elastic deformation temporary change in size or shape of a solid due to an applied force (stress); when force is removed the solid returns to its original size and shape

elasticity ability of a body that has been deformed by an applied force to return to its original shape when the force is removed

elastic modulus a measure of the stiffness of a solid material; defined as the ratio of stress to strain

electricity flow of energy due to the motion of electric charges; any physical effect that results from the existence of moving or stationary electric charges

electrode conductor (terminal) at which electricity passes from one medium into another; positive electrode is the *anode;* negative electrode is the *cathode*

electrolyte a chemical compound that, in an aqueous (water) solution, conducts an electric current

electromagnetic radiation (EMR) oscillating electric and magnetic fields that propagate at the speed of light. Includes in order of increasing frequency and energy: radio waves, radar waves, infrared (IR) radiation, visible light, ultraviolet radiation, X-rays, and gamma rays

electron (e) stable elementary particle with a unit negative electric charge (1.602×10^{-19} C). Electrons form an orbiting cloud, or shell, around the positively charged atomic nucleus and determine an atom's chemical properties

electron volt (eV) energy gained by an electron as it passes through a potential difference of one volt; one electron volt has an energy equivalence of 1.519×10^{-22} Btu = 1.602×10^{-19} J

element pure chemical substance indivisible into simpler substances by chemical means; all the atoms of an element have the same number of protons in the nucleus and the same number of orbiting electrons, although the number of neutrons in the nucleus may vary

elementary particle a fundamental constituent of matter; the basic atomic model suggests three elementary particles: the proton, neutron, and electron. *See also* **fundamental particle**

endothermic reaction chemical reaction requiring an input of energy to take place. *Compare* **exothermic reaction**

energy (E) capacity to do work; appears in many different forms, such as mechanical, thermal, electrical, chemical, and nuclear

entropy (S) measure of disorder within a system; as entropy increases, energy becomes less available to perform useful work

evaporation physical process by which a liquid is transformed into a gas (vapor) at a temperature below the boiling point of the liquid. *Compare with* **sublimation**

excited state state of a molecule, atom, electron, or nucleus when it possesses more than its normal energy. *Compare with* **ground state**

exothermic reaction chemical reaction that releases energy as it takes place. *Compare with* **endothermic reaction**

fatigue weakening or deterioration of metal or other material that occurs under load, especially under repeated cyclic or continued loading

fermion general name scientists give to a particle that is a matter constituent. Fermions are characterized by spin in odd half-integer quantum units (namely, 1/2, 3/2, 5/2, etc.); quarks, leptons, and baryons are all fermions

fission (nuclear) splitting of the nucleus of a heavy atom into two lighter nuclei accompanied by the release of a large amount of energy as well as neutrons, X-rays, and gamma rays

flavor in the standard model, quantum number that distinguishes different types of quarks and leptons. *See also* **quark; lepton**

fluid mechanics scientific discipline that deals with the behavior of fluids (both gases and liquids) at rest (fluid statics) and in motion (fluid dynamics)

foot-pound (force) ($ft\text{-}lb_{force}$) unit of work in American customary system of units; 1 $ft\text{-}lb_{force} = 1.3558$ J

force (F) the cause of the acceleration of material objects as measured by the rate of change of momentum produced on a free body. Force is a vector quantity mathematically expressed by Newton's second law of motion: force = mass × acceleration

freezing point the temperature at which a substance experiences a change from the liquid state to the solid state at a specified pressure; at this temperature, the solid and liquid states of a substance can coexist in equilibrium. *Synonymous with* **melting point**

fundamental particle particle with no internal substructure; in the standard model, any of the six types of quarks or six types of leptons and their antiparticles. Scientists postulate that all other particles are made from a combination of quarks and leptons. *See also* **elementary particle**

manufacturing process of transforming raw material(s) into a finished product, especially in large quantities

mass (m) property that describes how much material makes up an object and gives rise to an object's inertia

mass number *See* **relative atomic mass**

mass spectrometer instrument that measures relative atomic masses and relative abundances of isotopes

material tangible substance (chemical, biological, or mixed) that goes into the makeup of a physical object

mechanics branch of physics that deals with the motions of objects

melting point temperature at which a substance experiences a change from the solid state to the liquid state at a specified pressure; at this temperature, the solid and liquid states of a substance can coexist in equilibrium. *Synonymous with* **freezing point**

metallic bond chemical bond created as many atoms of a metallic substance share the same electrons

meter (m) fundamental SI unit of length; 1 meter = 3.281 feet. British spelling *metre*

metric system *See* **SI unit system**

metrology science of dimensional measurement; sometimes includes the science of weighing

microwave (radiation) comparatively short-wavelength electromagnetic (EM) wave in the radio frequency portion of the EM spectrum

mirror matter *See* **antimatter**

mixture a combination of two or more substances, each of which retains its own chemical identity

molarity (M) concentration of a solution expressed as moles of solute per kilogram of solvent

mole (mol) SI unit of the amount of a substance; defined as the amount of substance that contains as many elementary units as there are atoms in 0.012 kilograms of carbon-12, a quantity known as Avogadro's number (N_A), which has a value of about 6.022×10^{23} molecules/mole

molecule smallest amount of a substance that retains the chemical properties of the substance; held together by chemical bonds, a molecule can consist of identical atoms or different types of atoms

monomer substance of relatively low molecular mass; any of the small molecules that are linked together by covalent bonds to form a polymer

natural material material found in nature, such as wood, stone, gases, and clay

neutrino (ν) lepton with no electric charge and extremely low (if not zero) mass; three known types of neutrinos are the electron neutrino (ν_e), the muon neutrino (ν_μ), and the tau neutrino (ν_τ). *See also* **lepton**

neutron (n) an uncharged elementary particle found in the nucleus of all atoms except ordinary hydrogen. Within the standard model, the neutron is a baryon with zero electric charge consisting of two down (d) quarks and one up (u) quark. *See also* **standard model**

newton (N) The SI unit of force; 1 N = 0.2248 lbf

nuclear chain reaction occurs when a fissionable nuclide (such as plutonium-239) absorbs a neutron, splits (or fissions), and releases several neutrons along with energy. A fission chain reaction is self-sustaining when (on average) at least one released neutron per fission event survives to create another fission reaction

nuclear energy energy released by a nuclear reaction (fission or fusion) or by radioactive decay

nuclear radiation particle and electromagnetic radiation emitted from atomic nuclei as a result of various nuclear processes, such as radioactive decay and fission

nuclear reaction reaction involving a change in an atomic nucleus, such as fission, fusion, neutron capture, or radioactive decay

nuclear reactor device in which a fission chain reaction can be initiated, maintained, and controlled

nuclear weapon precisely engineered device that releases nuclear energy in an explosive manner as a result of nuclear reactions involving fission, fusion, or both

nucleon constituent of an atomic nucleus; a proton or a neutron

nucleus (plural: nuclei) small, positively charged central region of an atom that contains essentially all of its mass. All nuclei contain both protons and neutrons except the nucleus of ordinary hydrogen, which consists of a single proton

nuclide general term applicable to all atomic (isotopic) forms of all the elements; nuclides are distinguished by their atomic number, relative mass number (atomic mass), and energy state

ohm (Ω) SI unit of electrical resistance

oxidation chemical reaction in which oxygen combines with another substance, and the substance experiences one of three processes: (1) the gaining of oxygen, (2) the loss of hydrogen, or (3) the loss of electrons. In these reactions, the substance being "oxidized" loses electrons and forms positive ions. *Compare with* **reduction**

oxidation-reduction (redox) reaction chemical reaction in which electrons are transferred between species or in which atoms change oxidation number

particle minute constituent of matter, generally one with a measurable mass

pascal (Pa) SI unit of pressure; 1 Pa = 1 N/m^2 = 0.000145 psi

Pascal's principle when an enclosed (static) fluid experiences an increase in pressure, the increase is transmitted throughout the fluid; the physical principle behind all hydraulic systems

Pauli exclusion principle postulate that no two electrons in an atom can occupy the same quantum state at the same time; also applies to protons and neutrons

perfect fluid hypothesized fluid primarily characterized by a lack of viscosity and usually by incompressibility

perfect gas law *See* **ideal gas law**

periodic table list of all the known elements, arranged in rows (periods) in order of increasing atomic numbers and columns (groups) by similar physical and chemical characteristics

phase one of several different homogeneous materials present in a portion of matter under study; the set of states of a large-scale (macroscopic) physical system having relatively uniform physical properties and chemical composition

phase transition *See* **change of state**

photon A unit (or particle) of electromagnetic radiation that carries a quantum (packet) of energy that is characteristic of the particular radiation. Photons travel at the speed of light and have an effective momentum, but no mass or electrical charge. In the standard model, a photon is the carrier particle of electromagnetic radiation

photovoltaic cell *See* **solar cell**

physical property characteristic quality of a substance that can be measured or demonstrated without changing the composition or chemical identity of the substance, such as temperature and density. *Compare with* **chemical property**

Planck's constant (h) fundamental physical constant describing the extent to which quantum mechanical behavior influences nature. Equals the ratio of a photon's energy (E) to its frequency (ν), namely: $h = E/\nu = 6.626 \times 10^{-34}$ J-s (6.282×10^{-37} Btu-s). *See also* **uncertainty principle**

plasma electrically neutral gaseous mixture of positive and negative ions; called the fourth state of matter

plastic deformation permanent change in size or shape of a solid due to an applied force (stress)

plasticity tendency of a loaded body to assume a (deformed) state other than its original state when the load is removed

plastics synthesized family of organic (mainly hydrocarbon) polymer materials used in nearly every aspect of modern life

pneumatic operated, moved, or effected by a pressurized gas (typically air) that is used to transmit energy

polymer very large molecule consisting of a number of smaller molecules linked together repeatedly by covalent bonds, thereby forming long chains

positron (e^+ or β^+) elementary antimatter particle with the mass of an electron but charged positively

pound-force (lbf) basic unit of force in the American customary system; 1 lbf = 4.448 N

pound-mass (lbm) basic unit of mass in the American customary system; 1 lbm = 0.4536 kg

power rate with respect to time at which work is done or energy is transformed or transferred to another location; 1 hp = 550 ft-lb_{force}/s = 746 W

pressure (P) the normal component of force per unit area exerted by a fluid on a boundary; 1 psi = 6,895 Pa

product substance produced by or resulting from a chemical reaction

proton (p) stable elementary particle with a single positive charge. In the the standard model, the proton is a baryon with an electric charge of +1; it consists of two up (u) quarks and one down (d) quark. *See also* **standard model**

quantum mechanics branch of physics that deals with matter and energy on a very small scale; physical quantities are restricted to discrete values and energy to discrete packets called quanta

quark fundamental matter particle that experiences strong-force interactions. The six flavors of quarks in order of increasing mass are up (u), down (d), strange (s), charm (c), bottom (b), and top (t)

radiation heat transfer The transfer of heat by electromagnetic radiation that arises due to the temperature of a body; can takes place in and through a vacuum

radioactive isotope unstable isotope of an element that decays or disintegrates spontaneously, emitting nuclear radiation; also called radioisotope

radioactivity spontaneous decay of an unstable atomic nucleus, usually accompanied by the emission of nuclear radiation, such as alpha particles, beta particles, gamma rays, or neutrons

radio frequency (RF) a frequency at which electromagnetic radiation is useful for communication purposes; specifically, a frequency above 10,000 hertz (Hz) and below 3×10^{11}Hz

rankine (R) American customary unit of absolute temperature. *See also* **kelvin (K)**

reactant original substance or initial material in a chemical reaction

reduction portion of an oxidation-reduction (redox) reaction in which there is a gain of electrons, a gain in hydrogen, or a loss of oxygen. *See also* **oxidation-reduction (redox) reaction**

relative atomic mass (A) total number of protons and neutrons (nucleons) in the nucleus of an atom. Previously called *atomic mass* or *atomic mass number. See also* **atomic mass unit**

residual electromagnetic effect force between electrically neutral atoms that leads to the formation of molecules

residual strong interaction interaction responsible for the nuclear binding force—that is, the strong force holding hadrons (protons and neutrons) together in the atomic nucleus. *See also* **strong force**

resilience property of a material that enables it to return to its original shape and size after deformation

resistance (R) the ratio of the voltage (V) across a conductor to the electric current (I) flowing through it

scientific notation A method of expressing powers of 10 that greatly simplifies writing large numbers; for example, $3 \times 10^6 = 3,000,000$

SI unit system international system of units (the metric system), based upon the meter (m), kilogram (kg), and second (s) as the fundamental units of length, mass, and time, respectively

solar cell (photovoltaic cell) a semiconductor direct energy conversion device that transforms sunlight into electric energy

solid state of matter characterized by a three-dimensional regularity of structure; a solid is relatively incompressible, maintains a fixed volume, and has a definitive shape

solution When scientists dissolve a substance in a pure liquid, they refer to the dissolved substance as the *solute* and the host pure liquid as the *solvent.* They call the resulting intimate mixture the solution

spectroscopy study of spectral lines from various atoms and molecules; emission spectroscopy infers the material composition of the objects that emitted the light; absorption spectroscopy infers the composition of the intervening medium

speed of light *(c)* speed at which electromagnetic radiation moves through a vacuum; regarded as a universal constant equal to 186,283.397 mi/s (299,792.458 km/s)

stable isotope isotope that does not undergo radioactive decay

standard model contemporary theory of matter, consisting of 12 fundamental particles (six quarks and six leptons), their respective antiparticles, and four force carriers (gluons, photons, W bosons, and Z bosons)

state of matter form of matter having physical properties that are quantitatively and qualitatively different from other states of matter; the three more common states on Earth are solid, liquid, and gas

steady state condition of a physical system in which parameters of importance (fluid velocity, temperature, pressure, etc.) do not vary significantly with time

strain the change in the shape or dimensions (volume) of an object due to applied forces; longitudinal, volume, and shear are the three basic types of strain

stress applied force per unit area that causes an object to deform (experience strain); the three basic types of stress are compressive (or tensile) stress, hydrostatic pressure, and shear stress

string theory theory of quantum gravity that incorporates Einstein's general relativity with quantum mechanics in an effort to explain space-time phenomena on the smallest imaginable scales; vibrations of incredibly tiny stringlike structures form quarks and leptons

strong force In the standard model, the fundamental force between quarks and gluons that makes them combine to form hadrons, such as protons and neutrons; also holds hadrons together in a nucleus. *See also* **standard model**

subatomic particle any particle that is small compared to the size of an atom

sublimation direct transition of a material from the solid state to the gaseous (vapor) state without passing through the liquid phase. *Compare* **deposition**

superconductivity the ability of a material to conduct electricity without resistance at a temperature above absolute zero

temperature (T) thermodynamic property that serves as a macroscopic measure of atomic and molecular motions within a substance; heat naturally flows from regions of higher temperature to regions of lower temperature

tension condition when applied external forces pull atoms of a material farther apart. *Compare* **compression**

thermal conductivity (k) intrinsic physical property of a substance; a material's ability to conduct heat as a consequence of molecular motion

thermodynamics branch of science that treats the relationships between heat and energy, especially mechanical energy

thermodynamic system collection of matter and space with boundaries defined in such a way that energy transfer (as work and heat) from and to the system across these boundaries can be easily identified and analyzed

thermometer instrument or device for measuring temperature

toughness ability of a material (especially a metal) to absorb energy and deform plastically before fracturing

transmutation transformation of one chemical element into a different chemical element by a nuclear reaction or series of reactions

transuranic element (isotope) human-made element (isotope) beyond uranium on the periodic table

ultraviolet (UV) radiation portion of the electromagnetic spectrum that lies between visible light and X-rays

uncertainty principle Heisenberg's postulate that places quantum-level limits on how accurately a particle's momentum *(p)* and position *(x)* can be simultaneously measured. Planck's constant (h) expresses this uncertainty as $\Delta x \times \Delta p \geq h/2\pi$

U.S. customary system of units *See* **American customary system of units**

vacuum relative term used to indicate the absence of gas or a region in which there is a very low gas pressure

valence electron electron in the outermost shell of an atom

van der Waals force generally weak interatomic or intermolecular force caused by polarization of electrically neutral atoms or molecules

vapor gaseous state of a substance

velocity vector quantity describing the rate of change of position; expressed as length per unit of time

velocity of light *(c)* *See* **speed of light**

viscosity measure of the internal friction or flow resistance of a fluid when it is subjected to shear stress

volatile solid or liquid material that easily vaporizes; volatile material has a relatively high vapor pressure at normal temperatures

volt (V) SI unit of electric potential difference

volume (V) space occupied by a solid object or a mass of fluid (liquid or confined gas)

watt (W) SI unit of power (work per unit time); 1 W = 1 J/s = 0.00134 hp = 0.737 $ft\text{-}lb_{force}/s$

wavelength (λ) the mean distance between two adjacent maxima (or minima) of a wave

weak force fundamental force of nature responsible for various types of radioactive decay

weight *(w)* the force of gravity on a body; on Earth, product of the mass (m) of a body times the acceleration of gravity (*g*), namely $w = m \times g$

work (W) energy expended by a force acting though a distance. *Compare* **heat**

X-ray penetrating form of electromagnetic (EM) radiation that occurs on the EM spectrum between ultraviolet radiation and gamma rays

the early Greek philosophers to the 19th-century Russian chemist Dmitri Mendeleyev.

Thrower, Peter, and Thomas Mason. *Materials in Today's World.* 3rd ed. New York: McGraw-Hill Companies, 2007. Provides a readable introductory treatment of modern materials science, including biomaterials and nanomaterials.

Trefil, James, and Robert M. Hazen. *Physics Matters: An Introduction to Conceptual Physics.* New York: John Wiley & Sons, 2004. Highly-readable introductory college-level textbook that provides a good overview of physics from classical mechanics to relativity and cosmology. Laypersons will find the treatment of specific topics useful and comprehendible.

Zee, Anthony. *Quantum Field Theory in a Nutshell.* Princeton, N.J.: Princeton University Press, 2003. A reader-friendly treatment of the generally complex and profound physical concepts that constitute quantum field theory.

WEB SITES

To help enrich the content of this book and to make your investigation of matter more enjoyable, the following is a selective list of recommended Web sites. Many of the sites below will also lead to other interesting science-related locations on the Internet. Some sites provide unusual science learning opportunities (such as laboratory simulations) or in-depth educational resources.

American Chemical Society (ACS) is a congressionally chartered independent membership organization that represents professionals at all degree levels and in all fields of science involving chemistry. The ACS Web site includes educational resources for high school and college students. Available online. URL: http://portal.acs.org/portal/acs/corg/content. Accessed on February 12, 2010.

American Institute of Physics (AIP) is a not-for-profit corporation that promotes the advancement and diffusion of the knowledge of physics and its applications to human welfare. This Web site offers an enormous quantity of fascinating information about the history of physics from ancient Greece up to the present day. Available online. URL: http://www.aip.org/aip/. Accessed on February 12, 2010.

Chandra X-ray Observatory (CXO) is a space-based NASA astronomical observatory that observes the universe in the X-ray portion of the electromagnetic spectrum. This Web site contains contemporary information and educational materials about astronomy, astrophysics, and cosmology, including topics such as black holes, neutron stars, dark matter, and dark energy. Available online. URL: http://www.chandra.harvard.edu/. Accessed on February 12, 2010.

The ChemCollective is an online resource for learning about chemistry. Through simulations developed by the Department of Chemistry of Carnegie Mellon University (with funding from the National Science Foundation), a person gets the chance to safely mix chemicals without worrying about accidentally spilling them. Available online. URL: http://www.chemcollective.org/vlab/vlab.php. Accessed on February 12, 2010.

Chemical Heritage Foundation (CHF) maintains a rich and informative collection of materials that describe the history and heritage of the chemical and molecular sciences, technologies, and industries. Available online. URL: http://www.chemheritage.org/. Accessed on February 12, 2010.

Department of Defense (DOD) is responsible for maintaining armed forces of sufficient strength and technology to protect the United States and its citizens from all credible foreign threats. This Web site serves as an efficient access point to activities within the DOD, including those taking place within each of the individual armed services: the U.S. Army, U.S. Navy, U.S. Air Force, and U.S. Marines. As part of national security, the DOD sponsors a large amount of research and development, including activities in materials science, chemistry, physics, and nanotechnology. Available online. URL: http://www.defenselink.mil/. Accessed on February 12, 2010.

Department of Energy (DOE) is the single largest supporter of basic research in the physical sciences in the federal government of the United States. Topics found on this Web site include materials sciences, nanotechnology, energy sciences, chemical science, high-energy physics, and nuclear physics. The Web site also includes convenient links to all of the DOE's national laboratories. Available online. URL: http://energy.gov/. Accessed on February 12, 2010.

Fermi National Accelerator Laboratory (Fermilab) performs research that advances the understanding of the fundamental nature of matter and energy. Fermilab's Web site contains contemporary information about particle physics, the standard model, and the impact of particle physics on society. Available online. URL: http://www.fnal.gov/. Accessed on February 12, 2010.

Hubble Space Telescope (HST) is a space-based NASA observatory that has examined the universe in the (mainly) visible portion of the electromagnetic spectrum. This Web site contains contemporary information and educational materials about astronomy, astrophysics, and cosmology, including topics such as black holes, neutron stars, dark matter, and dark energy. Available online. URL: http://hubblesite.org/. Accessed on February 12, 2010.

Institute and Museum of the History of Science in Florence, Italy, offers a special collection of scientific instruments (some viewable online), including those used by Galileo Galilei. Available online. URL: http://www.imss.fi.it/. Accessed on February 12, 2010.

International Union of Pure and Applied Chemistry (IUPAC) is an international nongovernmental organization that fosters worldwide communications in the chemical sciences and in providing a common language for chemistry that unifies the industrial, academic, and public sectors. Available online. URL: http://www.iupac.org/. Accessed on February 12, 2010.

National Aeronautics and Space Administration (NASA) is the civilian space agency of the U.S. government and was created in 1958 by an act of Congress. NASA's overall mission is to direct, plan, and conduct American civilian (including scientific) aeronautical and space activities for peaceful purposes. Available online. URL: http://www.nasa.gov/. Accessed on February 12, 2010.

National Institute of Standards and Technology (NIST) is an agency of the U.S. Department of Commerce that was founded in 1901 as the nation's first federal physical science research laboratory. The NIST Web site includes contemporary information about many areas of science and engineering, including analytical chemistry, atomic and molecular physics, biometrics, chemical and crystal structure, chemical kinetics, chemistry, construction, environmental data, fire, fluids, material properties, physics, and thermodynamics. Available online. URL: http://www.nist.gov/index.html. Accessed on February 12, 2010.

National Oceanic and Atmospheric Administration (NOAA) was established in 1970 as an agency within the U.S. Department of Commerce to ensure the safety of the general public from atmospheric phenomena and to provide the public with an understanding of Earth's environment and resources. Available online. URL: http://www.noaa.gov/. Accessed on February 12, 2010.

NEWTON: Ask a Scientist is an electronic community for science, math, and computer science educators and students sponsored by the Argonne National Laboratory (ANL) and the U.S. Department of Energy's Office of Science Education. This Web site provides access to a fascinating list of questions and answers involving the following disciplines/topics: astronomy, biology, botany, chemistry, computer science, Earth science, engineering, environmental science, general science, materials science, mathematics, molecular biology, physics, veterinary, weather, and zoology. Available online. URL: http://www.newton.dep.anl.gov/archive.htm. Accessed on February 12, 2010.

Nobel Prizes in Chemistry and Physics. This Web site contains an enormous amount of information about all the Nobel Prizes awarded in physics and chemistry, as well as complementary technical information. Available online. URL: http://nobelprize.org/. Accessed on February 12, 2010.

Periodic Table of Elements. An informative online periodic table of the elements maintained by the Chemistry Division of the Department of Energy's Los Alamos National Laboratory (LANL). Available online. URL: http://periodic.lanl.gov/. Accessed on February 12, 2010.

PhET Interactive Simulations is an ongoing effort by the University of Colorado at Boulder (under National Science Foundation sponsorship) to provide a comprehensive collection of simulations to enhance science learning. The major science categories include physics, chemistry, Earth science, and biology. Available online. URL: http://phet.colorado.edu/index.php. Accessed on February 12, 2010.

ScienceNews is the online version of the magazine of the Society for Science and the Public. Provides insights into the latest scientific achievements and discoveries. Especially useful are the categories Atom and Cosmos, Environment, Matter and Energy, Molecules, and Science and Society. Available online. URL: http://www.sciencenews.org/. Accessed on February 12, 2010.

The Society on Social Implications of Technology (SSIT) of the Institute of Electrical and Electronics Engineers (IEEE) deals with such issues as the environmental, health, and safety implications of technology; engineering ethics; and the social issues related to telecommunications, information technology, and energy. Available online. URL: http://www.ieeessit.org/. Accessed on February 12, 2010.

Spitzer Space Telescope (SST) is a space-based NASA astronomical observatory that observes the universe in the infrared portion of the electromagnetic spectrum. This Web site contains contemporary information and educational materials about astronomy, astrophysics, and cosmology, including the infrared universe, star and planet formation, and infrared radiation. Available online. URL: http://www.spitzer.caltech.edu/. Accessed on February 12, 2010.

Thomas Jefferson National Accelerator Facility (Jefferson Lab) is a U.S. Department of Energy–sponsored laboratory that conducts basic research on the atomic nucleus at the quark level. The Web site includes basic information about the periodic table, particle physics, and quarks. Available online. URL: http://www.jlab.org/. Accessed on February 12, 2010.

United States Geological Survey (USGS) is the agency within the U.S. Department of the Interior that serves the nation by providing reliable scientific information needed to describe and understand Earth, minimize the loss of life and property from natural disasters, and manage water, biological, energy, and mineral resources. The USGS Web site is rich in science information, including the atmosphere and climate, Earth characteristics, ecology and environment, natural hazards, natural resources, oceans and coastlines, environmental issues, geologic processes, hydrologic processes, and water resources. Available online. URL: http://www.usgs.gov/. Accessed on February 12, 2010.

Index

Italic page numbers indicate illustrations.

A

absolute temperature 61
absolute temperature scales 27–28
absolute zero 10, 27
absorption 70
acceleration 38–39
accelerators 7, 175, 177
actinoid series 32, 65, 94
action and reaction 42
adamas 127
aerogel *157*, 157–158
aes Cyprium 100
Ag *See* silver (Ag)
aggregates 79, 121, 123
agriculture xiii–xiv, 80, 114
air, one of four classic elements xiv–xv, 6
air-entrained concrete 123
Air Force, U.S. *39*, 109, *109*
Al *See* aluminum (Al)
Albertus Magnus 169
alchemy xiv–xv, 169
Aldrin, Edwin "Buzz" 177
Alexander the Great 126
alkali metals 31, 93
alkaline earth metals 31, 94
allotropes 8, 124, 126, 127–128, 135
alloys xiv, 102, 103, 105, 106, 107
alloy steels 108
alumina (Al_2O_3) 111, 122
aluminum (Al) 29, 59, 90, 110–111
aluminum oxide (Al_2O_3) 90, 110–111
amalgams 106
amber 89
American customary unit system xvi, 21–22
amethyst 87–88, *88*
amorphous carbon 125, *129*, 129–130
amorphous silicon 141
amorphous solids 2, 52
ampere (A) 22
Ampère, André-Marie 66, 172
amphorae (amphora) 149
Anderson, Carl David 175
anemia 107
anisotropic substances 54
anthracite coal 130
antimatter 177
antimony 102
Apollo 11 177
Aqua Appia 119
aquaducts 118, 119–120
Aqua Maria 119–120
arches 117, 119
Archimedes of Syracuse 169
Archimedes screw 169
argentite (Ag_2S) 104
Aristotle 6
Armstrong, Neil 177
art and sculpture 113, 120–121
Aspdin, Joseph 122
astatine (At) 30
asteroids and comets 72
asthenosphere 78
Aston, Francis W. 175
atmospheric pressure 25
atomic bomb 176
atomic model of matter (atomism) *5*, 5–9, 16, *16*, 171
atomic nucleus 174
atomic number 30, 165–167
atomic weights xv
atoms 5, 7, 8, 169
Avogadro, Amedeo 172

B

Bacon, Roger 169
Baekeland, Leo 153, 174
Bakelite 153
ball clay 147
Bardeen, John 176
barter system 103
Bartholdi, Frédéric 101–102
basic (other) metals 95
basic units 22
basilicas 118
batteries 66, 67, 171
bauxite 111
Bayer, Karl Joseph 111
Bayer process 111
beach sand 139–141
Becher, Johann Joachim 170
becquerel (Bq) 22
Becquerel, Antoine-Henri 173
BECs (Bose-Einstein condensates) 177
bentonite 147, 148
Bernoulli, Daniel 171
beryl 88
Berzelius, Jöns Jacob 141, 172
Bessemer, Henry 108
Bessemer process 108

beta decay 175
Bethe, Hans Albrecht 176–177
big bang 9–13, 168
billiard balls 152–153
biochemistry 125
biological weapons 158
bituminous coal 130
Black, Joseph 62, 171
Blackbird (SR-71 aircraft) 109, *109*
blackbodies 61–62
black holes 45–49, *49*
Boltzmann, Ludwig 61
Borglum, Gutzon 86
borosilicate glass 145
Bose-Einstein condensates (BECs) 177
bosons 19
Boyle, Robert 170
Boyle, Willard 178
BQ (becquerel) 22
Brahe, Tycho 12
Brand, Hennig 170
brass 100, 102
Brattain, Walter 176
bricks 56, 63, 114, 148–149, 161
brittle materials 59–60
Broglie, Louis-Victor de 175
bromine 92, 96
bronze 100, 102
Bronze Age xiv, 100, 168
Btu 62
buckyballs 135, *135*, 177
bullion 98
Bullion Depository at Fort Knox, U.S. 98
Byzantine Empire 116

C

C *See* carbon (C)
calcium carbonate (calcite, limestone) 84, 118, 120, 121, 131
calcium oxide (CaO, quicklime) 117
caloric theory 60, 172
calx 117
candela (cd) 22
carat (ct) 87, 127
carbohydrates 131–132
carbon (C) 124–138
  amorphous carbon (coal, soot, charcoal, coke) 125, *129*, 129–130
  carbon cycle 130–132, 134
  carbon sequestration *133*, 133–134
  diamonds 8–9, 29, 79, 88, 124, 125–128, *126*
  fullerenes 125, 134–136, *135*
  general characteristics and applications 124–125
  graphite 8–9, 79, 124–125, 125–126, 128–129
  radiocarbon dating 137–138
carbonates/carbonate minerals *See* calcium carbonate
carbon cycle 130–132, 134
carbon-14 dating 125
carbon dioxide (CO_2) 64, 132–134, 171
carbonic acid 131
carbon sequestration *133*, 133–134
carbon steel 108
Carnot, Sadi 172
cave paintings 113
cd (candela) 22
celluloid 153, 173
cellulose 152
Celsius, Anders 27
Celsius scale 27
cement 117, 122–123
cementation 76
central processing units (CPUs) 143
ceramics 56, 123, 149
CERN (European Organization for Nuclear Research) 176, 177
cesium (Cs) 92
Chadwick, James 175
Challenger 53
Chandra X-ray Observatory (CXO) *12*, 46, 48
charcoal 125
chemical bonds 32–34
chemical properties of matter 20–21, 29–34
chemical weathering 76
Chicago Pile One (CP-1) 176
chips 144
chlorine 96
chromium 107
civilization, use of term 114
clarity 127
clay 147–149, 161
cleavage 79–80
CMB (cosmic microwave background) 176
Cn (copernicum) 178
coal 125, *129*, 129–130, 132
cobalt *7*
Cockcroft, John Douglas 175
coefficient of elasticity 58
coefficient of linear expansion 29
coke 125, 130
colonial period, U.S. 81
color (diamonds) 127
comets 72, 157–158
common clay 147
compaction 76
composites 150, 157–158
compression 56, 58, 59
compressive strength 52, 123
Compton, Arthur Holly 175
concrete 29, 59, 116–117, 122–123
condensed matter 4
conduction (conductors, conductivity) 60, 61, 62, 68, 93, 97, 100, 104, 142
conservation of energy 70, 172
Constantine the Great 116
Constantinople 116

constant temperature (isothermal) changes in state 64
convection 60, 61
copernicum (Cn) 178
Copernicus, Nicholas 40, 170
copper (Cu) *7*, 29, 68, 99–102, 104
copper carbonate ($CuCO_3$, malachite) 99, 101
copper chloride ($CuCl_2$) 101
core 73
Cornell, Eric A. 177
corrosion 91, *101*, 102, 110–111
corundum 84, 88
cosmic microwave background (CMB) 15–16, 176
cosmological constant 15
Coulomb, Charles-Augustin de 171
Coulomb's law 171
covalent bonds 33–34
CPUs (central processing units) 143
Cro-Magnon man xiii, 113, 168
crushed stone 121–122
crust 73, 74, 82–84
crystalline silicon 141
crystalline solids 2, *3*, 50–51, *51*
crystallography 174
ct (carat) 87, 127
Cu (copper) *7*, 29, 68, 99–102, 104
cubes 52
Cullinan, Thomas 128
Cullinan diamond 128
cuprite 100
cuprum 100
Curie, Marie and Pierre 174
Curl, Robert F., Jr. 134–135, 177
cut (diamonds) 127
cyclotrons 176
Cyprus 100

D

Dalton, John 171
dark energy 13–15, 17, 177
dark matter 13, *14*, 17
Davy, Humphry 172
decomposition 132
deformation 53, 56, 59
Democritus 5–6, 169
De Motu Corporum (On the motion of bodies, Newton) 42
density 23–24
Department of Energy, U.S. 133
deposition 64
derived units 22
diamagnetism 65
diamonds 8–9, 29, 79, 88, 124, 125–128, *126*
diffuse reflectors 70
digital revolution 69
dimension stone 120–121, *121*
dish detergent 156
dodecahedrons 52
dolomite 121
Dryden Flight Research Center 109
dry ice 64
ductility 59, 60, 92, 97–98, 103
DuPont Company 153, 176
dynamite 173
dynamos 172

E

Earth
  formation of 168
  surface and interior characteristics 72–74
earth, one of four classic elements xiv–xv, 6, 169
Edison, Thomas 68
Egypt 99, 115, *115*, 143–144
Einstein, Albert 11, 15, 47, 174
elasticity 54, *55*, 56–60, *58*
elastomers 54, 150, 152
electrical conductors 68, 142
electrical insulators 68, 142
electrical resistance 67
electric current (I) 67
electric generators 67
electricity 66–67
electric motors 67–68, 172
electromagnetic radiation (EM radiation) 61–62, 69
electromagnetism 64–69, 172
electronegativity 34
electrons 11, 173
electron shells 31
electrostatic attraction 33
electrum 103
elemental particles *See* particles
elements 160–161
  by atomic number 165–167
  families of 92–93
  formation of 9–13
  four classic Greek elements (earth, air, fire, water) xiv–xv, 6, 169
  metals 92–96
  periodic table xv, 29–32, 92–96, *164*, 164–165
$E = Mc^2$ 11
Empedocles 169
EM radiation (electromagnetic radiation) 61–62, 69
enthalpy of transformation 63
equations of state 25
equatorial radius 73
erosion 73, 75–76, 78, 131
ethics 136, 162–163
Euler, Leonhard 58
European Organization for Nuclear Research (CERN) 176, 177
European Space Agency 48
event horizon 47–48, *49*
exothermic processes 64
extrusive igneous rocks 74–75

F

F (fluorine) 96
fabric of space-time 44
Fahrenheit, Daniel Gabriel 26

lonsdaleite 127
Lower Paleolithic period xiii, 112–113
LSI (large-scale integration) 144
luster 82
Lydians 103, 169

M

magma 72–73
magnesium carbonate 134
magnetic dipole moments 65–66
magnetic flux 66
magnetite (Fe_3O_4) 107
magnets and magnetism 65–66, 82
malachite (copper carbonate) 99, 101
malleability 59, 92, 97–98
mantle 73–74
marble 118, 120
mass 1, 4, 13, 21–25
mass-energy formula ($E= Mc^2$) 11
materials science fundamentals 50–71
  amorphous solids 2, 52
  crystalline solids 2, *3*, 50–51, *51*
  elasticity 56–60
  electromagnetic properties 64–69
  heat transfer and thermodynamics 60–64
  materials processing in space 53, *53*
  optical properties 69–71
  shape 52
  strength 52, 54–56, *55*
matrix material 159
matter
  atomic model of *5*, 5–9
  concept of 1–5
  condensed v. uncondensed matter 4
  physical properties *See* physical properties of matter
  states of xii, *2*, 2–5, 63–64
  taking a close look at 16, *16*
Maxwell, James Clerk 67–68, 173
Mayer, Julius Robert von 172
mechanical equivalence of heat 173
mechanical weathering 76
melting point 63, 64
Mendeleev, Dmitri xv, 29–30, 173
Menkaure (pharaoh) 115–116
mercury (Hg) 106
mercury iodide 53, *53*
Mesolithic period (Middle Stone Age) xiii, 113
Mesopotamia 103
metabolism 132
metal fatigue 60
metallic bonding 34
metalloids 95
metals 90–111
  aluminum (Al) 110–111
  civilization and xiv, 161–162
  copper (Cu) 99–102
  general characteristics 90–96
  gold *97*, 97–99
  iron (Fe) 106–108
  lead (Pb) 105–106
  mercury (Hg) 106
  and modern technology 108
  silver (Ag) 103–104
  tin (Sn) 102–103
  titanium (Ti) 108–110
  type metal 105, 106
  used in antiquity 96–97
metal triads 94
metamorphic rocks (metamorphism) 74, 78–79, 118
meteorites 106, 107, 131
meter (m) 22
Michelangelo Buonarotti 120–121
Michell, John 46
Micrographia (Hooke) 170
Middle East 99–100, 103, 107
Middle Paleolithic period xiii, 112, 113
Middle Stone Age (Mesolithic period) xiii, 113
Milky Way galaxy 48
mille passuum 119
millesimal fineness system 98
minerals 79–84
  characteristics 79–82
  common 82–84
  definition 79
  sodium chloride (NaCl, table salt) 79, 80–81
Mint, U.S. 98
missing mass 13
modern atomic theory 8, 16, *16*
Mohs, Friedrich 80–82
Mohs scale of mineral hardness 81–82
mole (mol) 22
molecules 16, *16*
money, invention of 103
monomers 150
Moon 39–40
mortar 117
Moseley, Henry Gwyn Jeffreys 174
Mount Rushmore National Memorial 85–86

N

N (newton) 22
N_2 (nitrogen) 171
nanometer (nm) 22
nanotechnology 136, 162
nanotubes 136
NASA 15–16, 39–40, 46, 48, *53*, 109, *135*, 157–158, 159
  Chandra X-ray Observatory (CXO) *12*
National Institute of Standards and Technology (NIST) *7*
natural gemstones 87

Neanderthals xiii, 113, 168
Neolithic Revolution (New Stone Age) xiii–xiv, 80, 96, 113–114, 161, 168
neptunium (Np) 30
neutrinos 175
neutron capture process 11–12
newton (N) 22
Newton, Isaac 1, 40–43, 170
Nile River Valley 99
nitrogen (N_2) 171
nm (nanometer) 22
Nobel, Alfred 173
noble gases 31, 96
nonbaryonic matter 13
nonmetals 31, 95–96
nonrotating black holes 48, *49*
Noyce, Robert N. 144
Np (neptunium) 30
n-type semiconductors 142, 146–147
nuclear fission 176
nuclear reactors 176
nucleosynthesis 9
nylon 153, 176

O

O (oxygen) 24
obsidian 52
ocean sequestration 133–134
octahedrons 52
ohm (Ω) 67
Ohm, George Simon 66–67, 172
Ohm's law 67
Old Stone Age (Paleolithic period) xiii, 112–113
"On the Electrodynamics of Moving Bodies" (Einstein) 11
On the Revolution of Celestial Spheres (Copernicus) 170
opaque materials 70, 82
optics and optical devices 69–71
ordinary solid matter xiii, 17
ores 84, 107
Ørsted, Hans Christian 66, 172
osmium (Os) 24
other (basic) metals 95
oxidation 90–91, 102, 129
oxide minerals 83, 84
oxides 90
oxygen (O) 24

P

P (pressure) 25–26
Pa (pascal) 25
Pacific Ocean, Great Pacific Garbage Patch 154–155
Paleolithic period (Old Stone Age) xiii, 112–113
Pantheon 117
paramagnetism 65
particle accelerators 175, 177
particle board 159
particles xv–xvi, 17–19
particle-wave duality of matter 175
pascal (Pa) 25
Pascal, Blaise 25
patina 100–101, *101*
Pb (lead) 29, 105–106
pearls 89
peat 129
pencil "lead" 125
pentagons 52
periodic table of the elements xv, 29–32, 92–96, *164*, 164–165
permeability 140
pewter 102
phase changes *2*, 2–5
philosopher's stone xv
phlogiston theory 60, 170
photons 19
photosynthesis 131–132
photovoltaic (PV) converter cells 146
physical/biological carbon cycle 130, 131–134
physical properties of matter 20–34
  chemical properties 29–34
  defining 20–21
  interrelationship of 24–25
  mass 21–25
  pressure (P) 25–26
  temperature 26–29
pig iron 108
Ping-Pong balls 153
Pioneer 10 177
plain carbon steels 108
Planck, Max 174
planetary motion 41–42
plastics 54–55, 150–151, 152–155, 174
plates 74
plate tectonics 73, 74
platinum 98
Platonic solids 52
plumbing 106
plumbum 106
plutonism 74
plutonium (Pu) 30
plywood 159
Pm (promethium) 30
p-n junction 147
points 87
polar covalent bonds 34
pollution, plastic 154
polyethylene 125, 153
polygons 52
polyhedrals 52
polymers 54–55, 150–152, *151*
polymorphs 79
polypropylene 153
polystyrene 59, 153
polyvinyl chloride (PVC) 153–154
polyvinylidene chloride (PVDC) 154
porcelain 56, 149
porosity 140
portland cement 122
positrons 175
posttransition metals 94
potassium (K) 172
pottery xiv, 56, 114, 147, *148*, 148–149, 161

pound-force (lbf) 21–22
pound-mass (lbm) 21–22
Prandtl, Ludwig 174
precious stones 87
Premier diamond fields 128
pressure (P) 25–26
pressure differentials 25–26
Principia, The (Newton) 40, 42–43, 170–171
Principles of Chemistry (Mendeleev) 29–30, 173
printing press 105, 106, 169–170
promethium (Pm) 30
protoliths 78–79
p-type semiconductors 69, 142, 146–147
PVC (polyvinyl chloride) 153–154
PVDC (polyvinylidene chloride) 154
pyramids of Giza 115, *115*
pyrite 84

Q

quanta 174
quantum theory 174
quarks xv–xvi, 17
quartz (silicon dioxide, silica) *3*, 82–84, 88, 122, 139, 141
quicksilver 106

R

radian (rad) 22
radiation 60, 61, 69
radioactive decay 137, 174
radioactivity 173
radiocarbon dating 137–138
radio waves 173
Rankine scale (R) 10, 27–28
rapid neutron capture process (r-process) 12
rare earth elements 95
RCC (reinforced carbon-carbon) 159
reflection 70
Reflections on the Motive Power of Fire (Carnot) 172
reinforced carbon-carbon (RCC) 159
reinforcing material 159
Reissner-Nordström black holes 48
relative temperature scales 27
resistance 38, 67
resistivity 142
respiration 132
roads 118–119, 121–122
Robinson, Jonah, Leroy "Doane" 86
rock cycle 72, 74–79, *75*
Roentgen, Wilhelm Conrad 173
Roman Empire xiv, 80–81, 97, 106, 116–120
rotating black holes 49, *49*
Royal Society 41, 43
r-process (rapid neutron capture process) 12
rubber bands 58
rust 92
Rutherford, Daniel 171
Rutherford, Ernest 174, 175
R = V/I 67

S

s (second) 22
S (siemen) 67
salt *See* sodium chloride (NaCl, table salt)
salt mines 81
sand 139–141
sandblasting 141
sand castles 139, *140*
sandstone 120, 140
scanning tunneling microscopes *7*
Sceptical Chymist, The (Boyle) 170
Schrödinger, Erwin 175
Schwarzschild, Karl 47
Schwarzschild black hole 48
Schwarzschild radius 47
Scientific Revolution xiv, 1, 170
Scott, David R. 39–40
sculpture 120–121
Seaborg, Glenn 176
sea ice 28
seals *155*
second (s) 22
Second Industrial Revolution 66, 162
sedimentary rocks 74, 76, 78, 118
semiconductor devices 178
semiconductors 68–69, 141, 142
semiprecious stones 87–88
sensible heat 62–63
shape 52
shaving cream 156
shear 56
shear strength 52
sheen 82
shekels 103
Shockley, William 176
Si *See* silicon (Si)
siemen (S) 67
Siemens, Ernst Werner von 67
Sierra Nevada 77–78
silica (quartz, silicon dioxide) *3*, 82–84, 88, 122, 139, 141
silicon (Si)
  amorphous and crystalline forms 141–142
  clay and ceramics 147–149
  conductivity of 69, 141–146, 142–143
  glass 143–146
  resistivity 142
  silicones 143
  solar cells 146–147
silicon carbide (SiC) 143
silicon dioxide (quartz, silica) *3*, 82–84, 88, 139, 141
silicones 143
silver (Ag) 62–63, 68, 98, 103–104
silver bromide 104
silver nitrate 104

singularities 45, 47
SI units (International System of Units) xvi, 21, 22
skydiving 39, *39*
slate 120
slow neutron capture process (s-process) 11–12
Smalley, Richard E. 134–135, 177
small-scale integration (SSI) 144
smelting 99
soda lime glass 144
Soddy, Frederick 174
sodium 34
sodium chloride (NaCl, table salt) xiv, 33, *33*, 79, 80–81
soft matter 4, *4*, 150, 155–158
softwoods 152
solar cells 146–147
solar photovoltaic conversion 146
solar system 13
solder 102, 106
solids
  amorphous 2, 52
  crystalline 2, *3*, 50–51, *51*
  properties of 2
soot 125
space-time 44
special theory of relativity 11, 43, 174
specific gravity 79
specific heat 62–63
spectroscopy 70–71
specular reflectors 70
speed of light 43–45
Spitzer Space Telescope *135*
springs 56, 57–58
s-process (slow neutron capture process) 11–12
SSI (small-scale integration) 144
stained glass windows 145–146
stainless steel 107
standard model of fundamental particles xvi, 17–19, *18*
starch-polyester compositions 155
Stardust 157
stars 10–13, 160–161, 168
Statue of Liberty *101*, 101–102
steam power and steam engines 171
steel 59, 102, 107, 108
Stefan, Josef 61
Stefan-Boltzmann law 61
sterling silver 104
stiffness 54
Stone Age 112–116
stone building materials 112–123
  cement and concrete 122–123
  crushed stone 121–122
  dimension stone 120–121, *121*
  Roman Empire 116–120
  Stone Age 112–116
Stonehenge 114–115
strain 56–60, 58
strain hardening 59
Strassmann, Fritz 176
strength 52, 54–56, *55*
stresses 56–60
stress *versus* strain curve *57*
subbituminous coal 130
subduction 131
sublimation 64
sulfide minerals 84
Sumerians 99–100
Sun 161
superalloys 102
superheavy elements 30, 32, 95
supermassive black holes 46, 49
supernovas *12*, 15
synthetic diamonds 128
synthetic fibers 152, 153

T

taffy 58
tarnish 104
Tc (technetium) 30
technetium (Tc) 30
technical ceramics 56, 149
tectonics 73, 74
Teflon 154–155
temperature 26–29
temperature scales 10, 26–27
tensile strength 52
tension 56, 57–58
terminal velocity 39
terrestrial sequestration 133–134
tesla (T) 66
Tesla, Nikola 66, 68
tetrahedrons 52
Thales of Miletus 169
thermal conductivity/thermal conductors 62
thermal expansion 29
thermal insulators 62
thermal radiation 61
thermodynamics 60, 173
thermometers 26–27
thermonuclear devices 176
thermoplastic materials 54, 153
thermoscope 26
thermosetting materials 54, 153
Thomson, J. J. (Joseph John) 68, 173
Ti (titanium) 108–110
tin (Sn) 100, 102–103
titanium (Ti) 108–110
titanium dioxide (TiO_2) 110
tools 168–169
Torricelli, Evangelista 25, 170
transactinide elements 32
transistors 69, 143, 176
transition metals 31, 94
translucent materials 70, 82
transmission 70
transmutation xv
transparent materials (transparency) 70, 82
transuranium elements 32
Treatise of Elementary Chemistry (Lavoisier) 171
Treatise on Electricity and Magnetism (Maxwell) 173
triangles 52
Trinity explosion 176
troy ounces 98–99
Tycho's supernova *12*

Type Ia (carbon detonation)
 supernovas 15, 177
type metal 105, 106, 170

U

uncertainty principle 175
uncondensed matter 4
UNESCO. *See* United Nations
 Educational, Scientific and
 Cultural Organization
 (UNESCO)
uniform (general) corrosion 91
United Nations Educational,
 Scientific and Cultural
 Organization (UNESCO) 81
universe
  expansion of 13–15
  formation of 9–13
ununoctium (Uuo) 177
ununseptium (Uus) 178
Upper Paleolithic period xiii,
 113

V

V (voltage) 67
V (volume) 3, 22–23, *23*
vacuum pressure force 15
valence electrons 31
van der Waals forces 8–9, 126
vapor 3
vaults 117
verdigris 101
very large-scale integration
 (VLSI) 144
Via Appia 119
VLSI (very large-scale
 integration) 144
volcanism 72–73, 74–75
Volta, Alessandro 67, 171
voltage (V) 67
voltaic piles 66, 67, 171
volume (V) 3, 22–23, *23*

W

Walton, Ernest 175
water
  density 79
  ice 28, *28*
  one of four classic Greek
   elements xiv–xv, 6, 169
  phase change 2–3, 21, 34, 64
  temperature scales 27
  weathering and erosion 74–76
Watt, James 171
wave-particle duality of matter
 175
Wb (weber) 66
W bosons 19
weak gravitational lensing *14*
weathering 75–76, 78
weber (Wb) 66
Weber, Wilhelm Eduard 66
Wheeler, Archibald 46
whipped cream 156, *156*
whiteware 148, 149
Wieliczka Salt Mine 81
Wieman, Carl E. 177
*Wilkinson Microwave Anisotropy
 Probe* (*WMAP*) 16–17
Wilson, Robert Woodrow 176
wood 151–152, 159
wrought iron 108
wurtzite boron nitride (w-BN)
 127–128

X

XMM-Newton 48
X-ray crystallography 174
X-rays 173

Y

yield point 58
Yosemite National Park 77–78
Young's modulus of elasticity
 54, *55*, 58, 59

Z

Z bosons 19
zinc 100
Zosimos of Panoplis 169